BGE S1-S3
Mathematics & Numeracy

Third Level

Dr Helen Kelly
Kathleen McQuillan
Dr Alan Taylor

HODDER
GIBSON
AN HACHETTE UK COMPANY

The Publishers would like to thank the following for permission to reproduce copyright material.

Photo credits

p.16 © pamela_d_mcadams/stock.adobe.com; **p.17** © bryonyphotography/stock.adobe.com; **p.58** (top) © Anna Kucherova - Fotolia, (bottom) © Nitr - Fotolia; **p.72** © sergio27_120 – Fotolia; **p.73** © hotshotsworldwide - Fotolia.com; **p.104** © FreshPaint/ Shutterstock.com; **p.117** © LEDOMSTOCK/Shutterstock.com; **p.167** © James/stock.adobe.com; **p.175** © Nikolai Sorokin - Fotolia; **p.196** © Mmphotos/stock.adobe.com.

Acknowledgements

Every effort has been made to trace all copyright holders, but if any have been inadvertently overlooked, the Publishers will be pleased to make the necessary arrangements at the first opportunity.

Although every effort has been made to ensure that website addresses are correct at time of going to press, Hodder Gibson cannot be held responsible for the content of any website mentioned in this book. It is sometimes possible to find a relocated web page by typing in the address of the home page for a website in the URL window of your browser.

Hachette UK's policy is to use papers that are natural, renewable and recyclable products and made from wood grown in well-managed forests and other controlled sources. The logging and manufacturing processes are expected to conform to the environmental regulations of the country of origin.

Orders: please contact Hachette UK Distribution, Hely Hutchinson Centre, Milton Road, Didcot, Oxfordshire, OX11 7HH. Telephone: +44 (0)1235 827827. Email: education@hachette.co.uk. Lines are open from 9 a.m. to 5 p.m., Monday to Friday. You can also order through our website: www.hoddereducation.co.uk. If you have queries or questions that aren't about an order, you can contact us at hoddergibson@hodder.co.uk

First published in 2020 by
Hodder Gibson, an imprint of Hodder Education,
An Hachette UK Company
50 Frederick Street
Edinburgh, EH2 1EX
www.hoddergibson.co.uk

Impression number   5   4

Year   2024   2023

Cover photo © gudinny - stock.adobe.com

Illustrations by Aptara, Inc.

Typeset in Bliss Regular 12/14 by Aptara, Inc.

Printed and bound by CPI Group (UK) Ltd, Croydon, CR0 4YY

A catalogue record for this title is available from the British Library.

ISBN: 978 1 5104 7118 4

MIX
Paper | Supporting
responsible forestry
FSC
www.fsc.org
FSC™ C104740

SCOTLAND EXCEL

We are an approved supplier on the Scotland Excel framework.

Find us on your school's procurement system as **Hachette UK Distribution Ltd** or **Hodder & Stoughton Limited t/a Hodder Education**.

Contents

Introduction to BGE Mathematics

▶ How to get the most from this book

Mathematics is the richest language in the world. Learning mathematics is one of the most important things you can do to boost your brain power, now and throughout your life. Mathematicians understand, describe and influence the world around us.

This book covers all of the BGE Benchmarks for Mathematics at Third Level. The chapters take you on a journey which will both support and challenge you. Extension to Fourth Level material is included where appropriate. As you focus and work hard on each section you will gain knowledge, understanding and the ability to solve problems and communicate your solutions to others.

Working through this book will equip you to be the best mathematician you can be. Every topic is explained with rigour, taking no short cuts, and has an abundance of practice and challenge to suit every learner. The book is ambitious, encouraging you to aim high and build the skills you need for future success.

The book is enjoyable and easy to read. Every topic includes a clear, 'straight-to-the-point' method set out using bullet points in concise and simple language. It is full of worked examples and helpful hints and contains proven methods for mastering difficult concepts. This book has been designed both as a classroom aid and for your personal study.

≫ Build your skills and understanding

To get the most from this book, read the explanations and think about the ideas.
Follow the worked examples, paying close attention to the details.

> Each lesson starts with an explanation and memorable method to which you can refer.

> Each lesson has planned and progressive worked examples which prepare you to tackle the problem set.

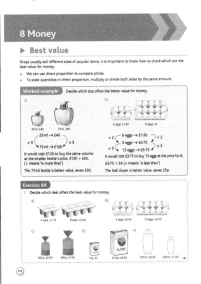

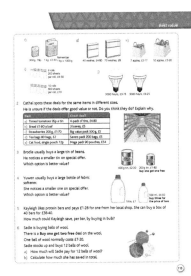

> Work through the exercise to practise and consolidate your skills. There are solutions at the end of the book, so check your work as you go to boost your confidence.

> Each exercise gets progressively harder. Try to finish the exercises and test your skills with the most challenging problems.

> Every chapter has a check-up exercise. Use these to bring your ideas together and boost your memory of key methods.

We hope that you enjoy using this book as much as we have enjoyed creating the content, questions and activities for you!

1 Number skills

▶ Addition and subtraction of decimals

When you are adding and subtracting decimals it is very important to take care with place value.

- Write the calculation out vertically, lining up the decimal points.
- Treat any columns with no digits in them as if they have a zero in them.

Worked examples

1. $0.2 + 0.7$

$$
\begin{array}{r}
0.2 \\
+ \ 0.7 \\
\hline
0.9
\end{array}
$$

2. $4.18 + 3.9$

$$
\begin{array}{r}
4.18 \\
+ \ 3.90 \\
\hline
8.08
\end{array}
$$

3. $16.85 + 0.467$

$$
\begin{array}{r}
16.850 \\
+ \ 0.467 \\
\hline
17.317
\end{array}
$$

4. $22 + 0.536 + 8.99$

$$
\begin{array}{r}
22.000 \\
+ \ 0.536 \\
8.990 \\
\hline
31.526
\end{array}
$$

5. Angus completes a dance routine to raise money for charity.
   He is sponsored £4·50, £7·30 and £2·75 by friends.
   How much money did Angus raise for charity?

$$
\begin{array}{r}
4.50 \\
+ \ 7.30 \\
2.75 \\
\hline
£14.55
\end{array}
$$

Exercise 1A

1. Add:
   a) $0.6 + 0.3$
   b) $6.4 + 2.2$
   c) $5.14 + 4.35$
   d) $20.142 + 35.367$
   e) $61.463 + 23.215$
   f) $46.79 + 19.83$
   g) $0.276 + 9.847$
   h) $58.639 + 8.305$

2. A cereal bar costs £1·35 and a tub of fruit costs £1·80. How much would it cost in total to buy one of each?

3. A piece of ribbon is cut into three separate lengths of 14·6 cm, 62·3 cm and 15·8 cm.
   What length was the original piece of ribbon?

4. Calculate:
   a) $0.4 + 0.31$
   b) $1.7 + 6.23$
   c) $8.14 + 1.5$
   d) $16.75 + 9.6$
   e) $0.14 + 0.367$
   f) $0.463 + 1.21$
   g) $8.56 + 0.289$
   h) $16.603 + 5.91$
   i) $47.6 + 97.354$
   j) $12.988 + 0.07$
   k) $1.96 + 12.582$
   l) $0.99 + 0.999$

5. Find:
   a) $0.2 + 2.3 + 8.4$
   b) $4.92 + 5.13 + 6.74$
   c) $19.214 + 41.327 + 15.008$
   d) $10.132 + 6.87 + 0.4$
   e) $27.299 + 4.7 + 16.99$
   f) $12.8 + 70.203 + 234.91$

6. In a triathlon, athletes have to complete a 0·93 mile swim, a 24·8 mile cycle and a 6·2 mile run.
   How far must an athlete cover altogether to complete the triathlon?

7. In a gymnastics competition, four judges award Diarmuid 9·150, 12·050, 13·200 and 10·354 points.
   How many points did Diarmuid score in total?

● When subtracting more than one decimal number, it is easier to add these numbers together before doing a single subtraction at the end.

Worked examples

1 $2.56 - 1.39$

$$\begin{array}{r} 2.\overset{4}{\cancel{5}}{}^{1}6 \\ -\ 1.3\ 9 \\ \hline 1.1\ 7 \end{array}$$

2 $6.12 - 0.394$

$$\begin{array}{r} \overset{5}{\cancel{6}}.\overset{1}{\cancel{1}}\overset{1}{\cancel{2}}{}^{1}0 \\ -\ 0.3\ 9\ 4 \\ \hline 5.7\ 2\ 6 \end{array}$$

3 $9.2 - 4.175 - 3.62$

Add the numbers you want to subtract.

$$\begin{array}{r} 4.1\ 7\ 5 \\ +\ 3.6\ 2\ 0 \\ \hline 7.7\ 9\ 5 \end{array}$$

Now do the subtraction.

$$\begin{array}{r} \overset{8}{\cancel{9}}.\overset{1}{\cancel{2}}\overset{9}{\cancel{0}}{}^{1}0 \\ -\ 7.7\ 9\ 5 \\ \hline 1.4\ 0\ 5 \end{array}$$

Exercise 1B

1 Find:

a) $0.4 - 0.1$ b) $6.9 - 1.3$ c) $16.54 - 7.31$ d) $9.49 - 3.37$

e) $5.452 - 0.201$ f) $8.24 - 2.15$ g) $19.75 - 16.81$ h) $3.47 - 0.89$

i) $2.417 - 1.628$ j) $19.121 - 8.235$ k) $6.238 - 4.516$ l) $0.206 - 0.098$

2 At the start of a journey a car has a meter reading of 245·6 miles. At the end of the journey the meter reading was 283·9 miles.
How far was the journey?

3 Calculate:

a) $4.87 - 2.3$ b) $9.32 - 0.1$ c) $734.6 - 12.31$ d) $58.437 - 29.2$

e) $62.384 - 51.03$ f) $12.461 - 7.56$ g) $88.465 - 66.97$ h) $108.11 - 43.512$

i) $123.456 - 7.89$ j) $1.04 - 0.009$ k) $18.061 - 2.09$ l) $0.4 - 0.018$

4 Calculate:

a) $37.9 - 2.2 - 10.6$ b) $148.91 - 31.203 - 5.6$ c) $71.192 - 43.539 - 25.62$

5 A large smoothie costs £3·95 and a small one costs £2·50.
How much change would you get from £10 if you bought one of each?

6 Ayla buys a football for £7·60 and a rugby ball for £9·85.
How much change will she get from £20?

7 The table shows the results in a 100 m race.

a) How much faster was the runner in first place than the runner in second place?

b) Which two runners had the closest finishing times to each other?

c) How much slower was the runner in fourth place than the runner in second place?

Place	Time in seconds
First	11·69
Second	11·75
Third	11·87
Fourth	11·91

The runner in fifth place was one-tenth of a second behind the runner in fourth place.

d) How long did the fifth-place runner take to complete the race?

▶ Single-digit multiplication

You should already be able to recall your times tables up to 12 quickly. If not, practise daily until you can.

- Write the calculation out vertically and start at the right-hand side.
- If any step gives a two-digit number, write down the units digit and carry the tens digit to the next place-value column to the left.
- When multiplying a decimal, remember to include the decimal point in your answer.

Worked examples

1 47×3

```
    4 7
×  ₂3
  1 4 1
```

2 5×628

```
    6 2 8
×  ₁ ₄5
  3 1 4 0
```

3 0.8×4

```
    0·8
×  ₃4
   3·2
```

4 17.29×8

```
    1 7·2 9
×  ₅ ₂ ₇8
  1 3 8·3 2
```

5 9.795×6

```
    9·7 9 5
×  ₄ ₅ ₃6
  5 8·7 7 0
```

6 A model car weighs 3·545 kg.

What is the total weight of seven identical model cars?

```
    3·5 4 5
×  ₃ ₃ ₃7
  2 4·8 1 5 kg
```

Exercise 1C

1 Calculate:

a) 26×4 b) 39×8 c) 75×3 d) 6×92 e) 7×49

f) 491×2 g) 639×5 h) 6×354 i) 1729×5 j) 4×3082

2 In a youth tournament, each football team brings 9 players.

If there are 24 teams in the tournament, how many players are there altogether?

3 Dan buys six packets of seeds to plant in his garden. Each packet has 45 seeds in it.

If every seed grows into a plant, how many plants will Dan get?

4 A small carton of milk is 189 ml. How many millilitres will there be in eight cartons?

5 Find:

a) 0.7×2 b) 1.4×8 c) 23.7×5 d) 6×48.1 e) 0.68×3

f) 2.51×4 g) 19.44×7 h) 9×34.12 i) 4.193×2 j) 16.724×5

6 A mug holds 0·225 litres. What volume would eight mugs hold?

7 In mathematics, irrational numbers are numbers that cannot be written exactly as a fraction of two integers (positive and negative whole numbers). Here are some irrational numbers approximated to three decimal places:

$$\sqrt{2} = 1.414 \qquad e = 2.718 \qquad \pi = 3.142$$

Use these values to approximate:

a) $5 \times \sqrt{2}$

b) $4 \times e$

c) the sum of $6 \times \pi$, $9 \times e$ and $7 \times \sqrt{2}$

d) the difference between $2 \times e$ and $3 \times \sqrt{2}$

▶ Single-digit division

To divide by a single-digit number:

- Start at the left-hand side. Carry any remainder one place-value column to the right.
- If there is a remainder after dividing the units, write a decimal point and introduce a zero to carry the remainder along.
- Continue to introduce zeros until the division is complete or you have achieved the required accuracy.

Worked examples

1 $95 \div 5$

$$5 \overline{)9\,{}^4 5} \quad = 19$$

2 $258 \div 3$

$$3 \overline{)2\,{}^2 5\,{}^1 8} \quad = 86$$

3 $15.18 \div 6$

$$6 \overline{)1\,{}^1 5 . {}^3 1\,{}^1 8} \quad = 2.53$$

4 $49 \div 2$

$$2 \overline{)4\,9 . {}^1 0} \quad = 24.5$$

5 $57 \div 4$

$$4 \overline{)5\,{}^1 7 . {}^1 0\,{}^2 0} \quad = 14.25$$

6 Eight friends share £46 equally. How much does each friend get?

$$8 \overline{)4\,{}^4 6 . {}^6 0\,{}^4 0} \quad = 5.75$$

They each get £5·75.

Exercise 1D

1 Find:
   a) $36 \div 3$ b) $85 \div 5$ c) $270 \div 6$ d) $973 \div 7$ e) $2452 \div 4$

2 750 ml of fruit juice is shared equally into three glasses. How much juice is in each glass?

3 Calculate:
   a) $20.7 \div 3$ b) $14.4 \div 6$ c) $72.3 \div 3$ d) $3.57 \div 7$ e) $12.64 \div 8$

4 Seven identical pens lined up in a straight line have a combined length of 115·5 cm. What is the length of one pen?

5 Find:
   a) $74 \div 5$ b) $35 \div 2$ c) $62 \div 4$ d) $2.7 \div 6$ e) $4.1 \div 2$
   f) $33.2 \div 8$ g) $19.8 \div 5$ h) $52.7 \div 2$ i) $14 \div 8$ j) $19.71 \div 5$

6 Four colleagues share a prize fund of £115 equally. How much does each of them get?

7 Five identical loaves of bread costs £6·60. What is the price of one loaf?

8 It takes 1500 g of flour to make six batches of scones.
   a) How much flour is required to make one batch?
   Within each batch there are eight scones.
   b) How much flour is required to make one scone?

9 An empty shipping container weighs 3·8 tonnes.
   After five identical crates are loaded into the container it weighs 4·1 tonnes.
   What is the weight of one crate?

▶ Rounding and estimation

When rounding, consider the digit in the place after the one you want to round to.

- If the digit is 5 or above, round the preceding digit up. Otherwise, leave the preceding digit unchanged.
- Unless told otherwise, always round answers involving money that include decimals to the nearest penny (2 decimal places).

Worked examples   1   Round:

a) 9·25 to one decimal place

     9·2:5   → next digit 5
     = 9·3   → 2 rounds up to 3

b) 42·394 to two decimal places

     42·39:4   → next digit 4
     = 42·39   → 9 remains
                   unchanged

c) 714·3162 to three decimal places

     714·316:2   → next digit 2
     = 714·316   → 6 remains
                     unchanged

d) 4·9999 to 3 decimal places

     4·999:9   → next digit 9
     = 5·000   → 9 continues to round up until the 4
                → The zeros are required to write the
                    answer to 3 decimal places

2   Three friends attempt to share £40 equally between them.

How much do they each get and how much is left over?

$$3\overline{)4^10\cdot{}^10^10^10\ldots}$$
$$13\cdot3\;3\;3\ldots$$

Each friend gets £13·33 and there is 1p left over

(check 13·33 × 3 = £39·99)

Exercise 1E

1   Round to 1 decimal place:

   a)   4·82   b)   7·19   c)   27·45   d)   85·37   e)   523·261

   f)   405·83   g)   0·08   h)   2·036   i)   6·99   j)   10·008

2   Round to 2 decimal places:

   a)   5·217   b)   4·391   c)   20·643   d)   204·815   e)   3147·746

   f)   0·4751   g)   16·299   h)   0·0092   i)   0·1800   j)   2·998

3   Round to 3 decimal places:

   a)   1·2048   b)   9·0814   c)   12·7446   d)   305·8769   e)   627·0483

   f)   0·002183   g)   0·07495   h)   10·0808   i)   12·0095   j)   7·0004

4   An exchange rate is advertised as 1 euro = £1·124. Round this value to 2 decimal places.

5   What is the minimum number of digits required after the decimal point to round to 5 decimal places?

                                                                         →

6   Three friends go to a restaurant for dinner. Their meals cost £11, £12·50 and £14·80. A coffee costs £1·90 and they each have one.

   a)   What is the total price of their dinner, including coffee?

   b)   If they split the bill evenly, what is the smallest amount they each must pay?

   They decide to add on a tip of £5 before splitting the bill evenly.

   c)   What is the minimum amount they must now each pay to ensure the bill is paid in full?

It is useful to check that your answer is roughly what you expect by making an **estimate**.

● Round given values to ease calculations.

● Obtain an estimate and compare to the exact calculation.

Worked examples   1   Approximate:

a) $2·9 \times 8$

Rounding to the nearest whole number this is approximately

3×8

$= 24$

The exact answer is 23·2.

b) $4·981 + 2·064 + 3·51$

Rounding to one decimal place this is approximately

$$\begin{array}{r} 5·0 \\ 2·1 \\ + \quad 3·5 \\ \hline 10·6 \end{array}$$

The exact answer is 10·555.

2   Denise has £10. She estimates that she can buy 2 magazines at £3·99 each and one birthday card at £1·79. Is she correct?

This is under $4 + 4 + 2 = £10$ so yes, she is correct.

The exact value is £9·77 and she would get 23p change.

Exercise 1F

1   a)   By estimating each calculation, determine whether Mary's answer is probably correct or is definitely wrong.

   b)   For the answers you think Mary got wrong, find the exact value of the calculation. Is your answer close to your estimation?

Question	Mary's answer
$18·9 + 12 + 13·8$	44·7
$26·7 \times 4$	98·8
$23·193 + 16·82 - 2·58$	47·423
$611·8 \div 8$	76·475
$35·267 \times 9$	3174·03

2   Kamran has £20 to spend and he intends to buy some of these items for his young child:

   Nappies: £8·90 per pack   Baby wipes: £0·87 per packet   Shampoo: £2·15 per bottle

   Vests: £7·95 per pack   Bubble bath: £2·10 per bottle   Cotton wool: £1·40 per pack.

   a)   Use estimates of the prices of these items to determine whether he can buy:

      i)   One bottle of shampoo, one pack of nappies and one pack of vests

      ii)   Two packs of nappies, two packs of wipes and one pack of cotton wool

      iii)   Two packs of vests, one bottle of bubble bath and one bottle of shampoo.

   b)   Calculate the exact value of the three shopping lists in **a)**. Were your estimates correct?

3   Why is it better to overestimate money calculations than underestimate?

▶ Multiplication and division by multiples of 10

- Digits move left when numbers are multiplied by 10.
- Digits move right when numbers are divided by 10.
- When numbers are multiplied or divided by 10 the digits move one place, by 100 they move two places, by 1000 they move three places, etc.

Multiplying

Tens	Units	·	Tenths	Hundredths	Thousandths
1	3	·	0	5	7

Dividing

Worked examples

1   $13 \times 10 = 130$

2   $2.7 \times 10 = 27$

3   $2.09 \times 10 = 20.9$

4   $0.68 \times 100 = 68$

5   $93.0518 \times 1000 = 93051.8$

6   $1570 \div 10 = 157$

7   $261 \div 10 = 26.1$

8   $5 \div 100 = 0.05$

9   $31.54 \div 100 = 0.3154$

10   $2493.7 \div 1000 = 2.4937$

11   $16 \div 1000 = 0.016$

Exercise 1G

1   Calculate:
   a) 43×10
   b) 7.1×10
   c) 19.06×10
   d) 914.345×10
   e) 89×100
   f) 0.15×100
   g) 16.873×100
   h) 82.0951×100
   i) 613×1000
   j) 51.2×1000
   k) 0.327×1000
   l) 4.1009×1000

2   Find:
   a) $190 \div 10$
   b) $62 \div 10$
   c) $74.8 \div 10$
   d) $1.35 \div 10$
   e) $4269 \div 100$
   f) $663.4 \div 100$
   g) $47.61 \div 100$
   h) $7.624 \div 100$
   i) $12507 \div 1000$
   j) $428 \div 1000$
   k) $12.97 \div 1000$
   l) $1.025 \div 1000$

To multiply or divide by multiples of 10 other than 100, 1000, etc. take two steps.

- Decompose the multiple into 10, 100, 1000, etc. and another factor.
- Multiply or divide by both factors.

It does not matter which factor you multiply or divide by first but try to make the calculation easier by avoiding decimals if you can.

Worked examples

1   34×50

$\times 50 = \times 5 \times 10$

$$\begin{array}{r} 3\,4 \\ \times\ \ _2 5 \\ \hline 1\,7\,0 \end{array}$$

$170 \times 10 = 1700$

$34 \times 50 = 1700$

2   63.2×700

$\times 700 = \times 7 \times 100$

$$\begin{array}{r} 6\,3.2 \\ \times\ \ _{2\ 1}7 \\ \hline 4\,4\,2.4 \end{array}$$

$442.4 \times 100 = 44240$

$63.2 \times 700 = 44240$

3   400×18.321

$\times 400 = \times 4 \times 100$

$18.321 \times 100 = 1832.1$

$$\begin{array}{r} 1\,8\,3\,2.1 \\ \times\ \ _{3\ 1}4 \\ \hline 7\,3\,2\,8.4 \end{array}$$

$400 \times 18.321 = 7328.4$

3 Copy and complete:
   a) $\times\ 60 = \times\ \boxed{} \times 10$
   b) $\times\ 90 = \times\ 9 \times \boxed{}$
   c) $\times\ 800 = \times\ \boxed{} \times 100$
   d) $\times\ 300 = \times\ 3 \times \boxed{}$
   e) $\times\ 2000 = \times\ \boxed{} \times 1000$
   f) $\times\ 7000 = \times\ 7 \times \boxed{}$

4 Find:
   a) 18×30
   b) 42×50
   c) $1 \cdot 6 \times 20$
   d) $15 \cdot 9 \times 80$
   e) $6 \cdot 32 \times 40$
   f) $38 \cdot 26 \times 60$
   g) $6 \cdot 1 \times 200$
   h) $13 \cdot 825 \times 600$
   i) $8 \cdot 271 \times 500$
   j) $20 \cdot 367 \times 300$
   k) $12 \cdot 5 \times 7000$
   l) $62 \cdot 0351 \times 2000$

5 A box of eggs costs £2·15. What is the price of 30 boxes of eggs?

6 A school enterprise group are selling tickets for a charity concert.
   If one ticket costs £4·75 how much money will they raise if they sell 400 tickets?

Worked examples

1 $1260 \div 40$

$\div 40 = \div 4 \div 10$

$$4 \overline{)1^12\ 6^20}$$
$$3\ 1\ 5$$

$315 \div 10 = 31 \cdot 5$

$1260 \div 40 = 31 \cdot 5$

2 $20 \cdot 7 \div 900$

$\div 900 = \div 9 \div 100$

$$9\overline{)2^20 \cdot {}^27}$$
$$2 \cdot 3$$

$2 \cdot 3 \div 100 = 0 \cdot 023$

$20 \cdot 7 \div 900 = 0 \cdot 023$

3 $75 \div 2000$

$\div 2000 = \div 2 \div 1000$

$$2\overline{)7^15 \cdot {}^10}$$
$$3\ 7 \cdot 5$$

$37 \cdot 5 \div 1000 = 0 \cdot 0375$

$75 \div 2000 = 0 \cdot 0375$

7 Copy and complete:
   a) $\div 20 = \div 2 \div \boxed{}$
   b) $\div 50 = \div \boxed{} \div 10$
   c) $\div 600 = \div 6 \div \boxed{}$
   d) $\div 800 = \div \boxed{} \div 100$
   e) $\div 4000 = \div 4 \div \boxed{}$
   f) $\div 7000 = \div \boxed{} \div 1000$

8 Calculate:
   a) $640 \div 40$
   b) $130 \div 50$
   c) $72 \cdot 6 \div 20$
   d) $5 \cdot 4 \div 30$
   e) $192 \cdot 5 \div 70$
   f) $2460 \div 600$
   g) $1581 \div 300$
   h) $803 \cdot 6 \div 200$
   i) $742 \cdot 5 \div 900$
   j) $17 \cdot 5 \div 700$
   k) $1956 \div 3000$
   l) $329 \cdot 2 \div 4000$

9 An electrician requires twenty identical lengths of wire from a 1·4 metre roll.
   What length should he cut each piece of wire to make sure they are the same?

10 Alfie gets paid £343·80 for working a thirty-hour week. How much does he get paid per hour?

11 A bus company sell 400 all-day adult tickets in one day. They collected £1980 from these tickets.
   How much does an adult ticket with this bus company cost?

12 A gas company charges customers a 70p daily charge plus 3·1p per unit of gas used.
   a) How much was the bill for August if a household used 700 units of gas? Write your answer in pounds and pence.
   Another gas bill was for 900 units of gas and totalled £45·40.
   b) How many days was this bill for?

► Long multiplication

- Write the calculation vertically, aligning digits to the right-hand side.
- Multiply by the units digit of the multiplier and write this answer below the line.
- Write a zero in the right-hand column, below the first multiplication, then multiply by the tens digit of the multiplier.
- Continue for hundreds (writing two zeros), thousands (writing three zeros), etc.
- Add to obtain the final answer.

Worked examples

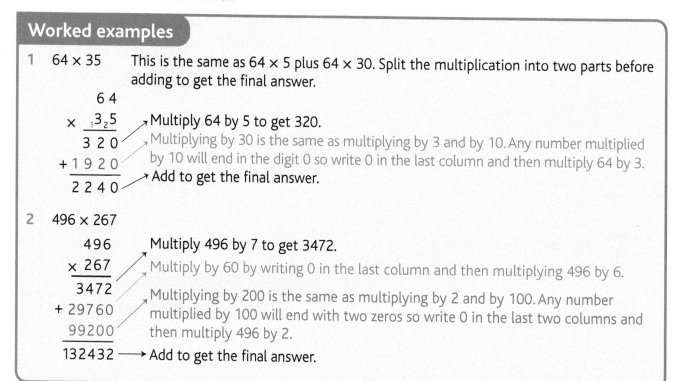

1   64 × 35   This is the same as 64 × 5 plus 64 × 30. Split the multiplication into two parts before adding to get the final answer.

```
        6 4
    ×  ₁3₂5      Multiply 64 by 5 to get 320.
      3 2 0      Multiplying by 30 is the same as multiplying by 3 and by 10. Any number multiplied
    + 1 9 2 0    by 10 will end in the digit 0 so write 0 in the last column and then multiply 64 by 3.
      2 2 4 0    Add to get the final answer.
```

2   496 × 267

```
        4 9 6    Multiply 496 by 7 to get 3472.
    ×   2 6 7    Multiply by 60 by writing 0 in the last column and then multiplying 496 by 6.
      3 4 7 2
    + 2 9 7 6 0  Multiplying by 200 is the same as multiplying by 2 and by 100. Any number
      9 9 2 0 0  multiplied by 100 will end with two zeros so write 0 in the last two columns and
                 then multiply 496 by 2.
    1 3 2 4 3 2  Add to get the final answer.
```

Exercise 1H

1   Find:

  a) 23 × 46   b) 57 × 29   c) 62 × 78   d) 55 × 42   e) 86 × 34

2   A bus seats 56 passengers. How many passengers will 18 buses seat?

3   It takes 72 screws to build a bed. How many screws are required to build 25 beds?

4   Jamie delivers 23 newspapers every day. How many papers does she deliver in July if she has no days off?

5   Calculate:

  a) 156 × 27   b) 218 × 59   c) 473 × 86   d) 683 × 469   e) 298 × 356

6   A return trip from Glasgow to Manchester is 428 miles. How far would 32 such return trips be?

7   A small glass can hold 125 ml of liquid. How much liquid would it take to fill 64 small glasses?

8   You have the digits 2, 5, 7 and 9 and the multiplication sign. You can put the digits together to make other numbers, for example 25, 79 and 257. If you must use each of them once only, what is the

  a) biggest number you can make   b) smallest number you can make?

- Follow the same method when multiplying a whole number by a decimal number. Take care to align the decimal points.

Worked examples

1 23·7 × 16

```
   23·7
 ×   16
  142·2
  237·0
  379·2
```

2 9·83 × 437

```
      9·83
 ×     437
    68·81
   294·90
  3932·00
  4295·71
```

3 16·382 × 5491

```
      16·382
 ×      5491
     16·382
   1474·380
   6552·800
  81910·000
  89953·562
```

4 A football ticket costs £28·95.

How much money will be raised if 2819 people attend a game?

```
      28·95
 ×     2819
    260·55
    289·50
  23160·00
  57900·00
  81610·05
```

£81610·05 will be raised.

Exercise 1I

1 Find:
   a) 3·7 × 24
   b) 62·3 × 98
   c) 9·65 × 41
   d) 18·17 × 56
   e) 9·6 × 241
   f) 57·3 × 682
   g) 8·52 × 733
   h) 5·61 × 472

2 Ayda is paid £9·85 per hour. How much will she be paid for working a 35-hour week?

3 Morris charges his customers £2·15 for a bag of compost. How much would 62 bags of compost cost?

4 A professional golf lesson costs £28·90. How much does the professional make for teaching 27 lessons in two weeks?

5 Calculate:
   a) 23·17 × 6831
   b) 42·168 × 3912
   c) 2·184 × 2375
   d) 72·367 × 4106

6 It costs £8·15 to download an album. In the first hour after release, the album is downloaded 2536 times. How much money has the album made in the first hour?

7 A concert venue advertises three ticket types, as shown in the table.
   How much money would the venue make if it sold:

Ticket type	Cost
Standing	£39·85
Premium seating	£48·50
Standard seating	£41·15

   a) 1625 standing tickets
   b) 280 premium seating tickets
   c) 3139 standard seating tickets?
   These prices exclude a charity donation of 5p per ticket.
   d) Calculate the total takings for selling 1625 standing, 280 premium seating and 3139 standard seating tickets if the charity donation is added.

8 It costs £5·35 per week to subscribe to a digital edition of a well-known financial newspaper.
   How much will the newspaper make in one year if they have 8673 digital subscribers?

▶ Addition and subtraction of integers

- Add to or subtract from the first number given.

Worked examples

1 −2 + 5

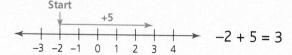

−2 + 5 = 3

2 −6 + 4

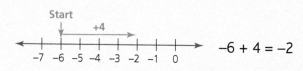

−6 + 4 = −2

3 1 − 4

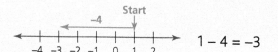

1 − 4 = −3

4 −2 − 3

−2 − 3 = −5

Exercise 1J

1 Find:

  a) −1 + 6   b) −3 + 9   c) −4 + 7   d) −2 + 10   e) −8 + 11

  f) −8 + 3   g) −4 + 1   h) −9 + 5   i) −10 + 4   j) −12 + 12

2 Jax gets into a lift in the basement which is level −2. He goes up 6 floors. What level does Jax get out at?

3 Calculate:

  a) 4 − 5   b) 3 − 7   c) 6 − 9   d) 5 − 10   e) 1 − 14

  f) −1 − 3   g) −5 − 2   h) −2 − 7   i) −3 − 10   j) −6 − 12

4 On a given day, the temperature in Glasgow is 13 °C and the temperature in Moscow is 21 degrees lower. What is the temperature in Moscow?

5 Find:

  a) −4 − 26   b) 8 − 32   c) −25 + 4   d) −30 − 12   e) −45 + 13

  f) −100 + 40   g) −204 − 51   h) 7 − 100   i) 18 − 70   j) 53 − 98

6 Scientists often use the Kelvin scale for temperature. Zero kelvin is approximately −273 °C and an increase of 1 kelvin is the same as an increase of 1 °C. What temperature is 14 kelvin in degrees Celsius?

7 In golf, a negative score means you are under par and a positive score means you are over par. If a player is under par, then they have taken fewer shots than a player over par.

  Matthew is 9 under par and Gerry takes 15 more shots than Matthew. What is Gerry's score?

Addition is **commutative**, which means you can add numbers in any order. Consider −2 and 5.

Adding them can be written as   −2 + 5 = 3   OR   5 + (−2) = 3.

The difference between them is 7 which can be written as 5 − (−2) = 7.

- Adding a negative number is the same as subtracting its **modulus** (positive value)
  5 + (−2) = 5 − 2 = 3.
- Subtracting a negative number is the same as adding its modulus 5 − (−2) = 5 + 2 = 7.

Worked examples

1 8 + (−5)
   = 8 − 5
   = 3

2 3 + (−9)
   = 3 − 9
   = −6

3 −4 + (−1)
   = −4 − 1
   = −5

4 5 − (−8)
   = 5 + 8
   = 13

5 −6 − (−10)
   = −6 + 10
   = 4

6 −12 − (−3)
   = −12 + 3
   = −9

Exercise 1K

1 Find:
   a) 7 + (−3)   b) 15 + (−6)   c) 23 + (−1)   d) 0 + (−2)   e) 1 + (−5)
   f) 4 + (−10)   g) 8 + (−12)   h) −1 + (−4)   i) −3 + (−6)   j) −2 + (−9)
   k) −4 + (−3)   l) −5 + (−10)   m) −9 + (−3)   n) −12 + (−7)   o) −25 + (−2)

2 Calculate:
   a) 4 − (−2)   b) 3 − (−1)   c) 8 − (−3)   d) 6 − (−5)   e) 12 − (−4)
   f) 20 − (−9)   g) 43 − (−12)   h) −3 − (−5)   i) −2 − (−7)   j) −6 − (−8)
   k) −10 − (−1)   l) −8 − (−3)   m) −12 − (−6)   n) −20 − (−10)   o) −30 − (−14)

3 Compute:
   a) −2 + (−7)   b) 3 − (−5)   c) −9 + (−3)   d) −11 − (−7)   e) 2 + (−9)
   f) −3 + (−10)   g) 4 + (−16)   h) −8 + (−7)   i) −10 − (−15)   j) 60 + (−80)
   k) −20 + (−42)   l) −30 − (−61)   m) 73 − (−24)   n) −41 − (−18)   o) −135 − (−30)

4 In each cloud, simplify to find the number that only appears once.

a)

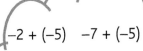

−2 + (−5)   −7 + (−5)

−9 − (−3)   −18 − (−6)

5 + (−12)   6 + (−4)

−1 + (−5)

b)

−1 − (−8)   −8 − (−2)

9 + (−2)   −5 − (−3)

3 + (−12)   2 + (−4)

−5 + (−1)

c)

−3 + (−2) −1

−4 + 3 −(−2)

−5 − (−4) + 2

2 + (−4) + (−2)

−8 + (−1) + 3

▶ Multiplication and division of integers

Consider 3×4. This is the same as 3 lots of 4 which can be written as $4 + 4 + 4 = 12$.

Now consider $3 \times (-4)$. This is 3 lots of (-4) which can be written as $-4 + (-4) + (-4) = -4 - 4 - 4 = -12$.

Look at the multiplication grid.

It also shows that $3 \times (-4) = -12$.

The pattern continues, in red, along the rows and down the columns for multiples of negative numbers.

From this, you can see that $-3 \times (-4) = 12$.

- A positive integer × a negative integer equals a negative integer.
- A negative integer × a positive integer equals a negative integer.
- A negative integer × a negative integer equals a positive integer.

×	3	2	1	0	−1	−2	−3	−4
3	9	6	3	0	−3	−6	−9	−12
2	6	4	2	0	−2	−4	−6	−8
1	3	2	1	0	−1	−2	−3	−4
0	0	0	0	0	0	0	0	0
−1	−3	−2	−1	0	1	2	3	4
−2	−6	−4	−2	0	2	4	6	8
−3	−9	−6	−3	0	3	6	9	12
−4	−12	−8	−4	0	4	8	12	16

Worked examples

1 $5 \times (-9)$
$= -45$

2 $(-3) \times 20$
$= -60$

3 $-9 \times (-14)$
$= 126$

4 $2 \times (-4) \times 9$ → multiply two integers together first
$= -8 \times 9$
$= -72$ → complete calculation

Exercise 1L

1 Calculate:
   a) $5 \times (-2)$ b) $8 \times (-4)$ c) $7 \times (-12)$ d) $3 \times (-50)$ e) -4×5
   f) -7×2 g) -3×22 h) -8×20 i) $-3 \times (-8)$ j) $-5 \times (-8)$
   k) $-9 \times (-4)$ l) $-6 \times (-13)$ m) $-10 \times (-9)$ n) $-32 \times (-9)$ o) $-83 \times (-5)$

2 Simplify:
   a) $-2 \times 3 \times 4$ b) $5 \times (-3) \times 2$ c) $4 \times (-2) \times (-9)$ d) $-7 \times (-3) \times 5$

3 Find the output value for each chain.

   a) Input [6] → × (−3) → + 7 → × (−5) → − 4 → × 2 → ( ) Output

   b) Input [−2] → × (−5) → + 6 → × (−1) → − (−6) → × 9 → ( ) Output

   c) Input [4] → × (−2) → + (−4) → × (−7) → + 3 → × 3 → ( ) Output

   d) Input [−8] → × (−7) → + (−2) → × (−6) → − (−9) → × 3 → ( ) Output

If $3 \times (-4) = -12$, then $\dfrac{-12}{3} = -4$ and $\dfrac{-12}{-4} = 3$.

- A positive integer ÷ a negative integer equals a negative integer.
- A negative integer ÷ a positive integer equals a negative integer.
- A negative integer ÷ a negative integer equals a positive integer.

Worked examples

1 $20 \div (-2)$
   $= -10$

2 $-42 \div 3$
   $= -14$

3 $-104 \div (-8)$
   $= 13$

4 $-258 \div 4$

$$\dfrac{-258}{4}$$

or

$$4\overline{)2^25^18 \cdot {}^20}$$ giving $6\,4 \cdot 5$

$$= -\dfrac{129}{2}$$

so the answer is $-64 \cdot 5$

When the division is more complicated, write as a fraction and simplify or do the division calculation using positive integers and then write the sign with your final answer.

Exercise 1M

1 Evaluate:

a) $12 \div (-6)$ b) $21 \div (-7)$ c) $56 \div (-8)$ d) $48 \div (-4)$ e) $-15 \div 5$

f) $-72 \div 9$ g) $-30 \div 6$ h) $-90 \div 3$ i) $-260 \div 10$ j) $105 \div (-5)$

k) $123 \div (-3)$ l) $-924 \div 7$ m) $-36 \div (-9)$ n) $-88 \div (-8)$ o) $-114 \div (-6)$

2 Evaluate (your answer will not be an integer):

a) $54 \div (-4)$ b) $-109 \div (-5)$ c) $-36 \div 10$ d) $-218 \div 8$

3 Find the value of x in each chain.

a)
Input 4 × (−2) + 10 × (−6) ÷ 3 + 1 x Output

b)
Input 30 ÷ (−5) + 2 × (−8) + 4 ÷ (−6) x Output

c)
Input 2 × 7 + (−4) × (−2) + 10 ÷ x 2 Output

d)
Input x + 3 × (−2) + 12 × (−6) ÷ (−4) 3 Output

4 Find two numbers which

a) when divided have a quotient of −6 and when added sum to 10

b) when divided have a quotient of −4 and have a difference of 40

c) when divided have a quotient of 2 and when added sum to −6

▶ Order of operations – BODMAS

Think about 5 + 3 × 2. Is the answer 8 × 2 = 16 or 5 + 6 = 11?

Now consider   These add to give 11p so 5 + 3 × 2 = 11.

● It is important that calculations are completed in the correct order. The mnemonic BODMAS can help:

Brackets   simplify inside any brackets first

Of

Division   } division and multiplication have the same priority
Multiplication

Addition   } addition and subtraction have the same priority
Subtraction

Worked examples

1 $20 - 3 \times 6$

$20 - 3 \times 6 \rightarrow$ multiply

$= 20 - 18 \quad \rightarrow$ subtract

$= 2$

2 $8 + \dfrac{1}{2}$ of $10 - 2$

$8 + \dfrac{1}{2}$ of $10 - 2 \rightarrow$ of

$= 8 + 5 - 2 \quad \rightarrow$ add and

$= 11 \qquad\qquad$ subtract

3 $(4 + 9) \times 2 + 27 \div 9$

$(4 + 9) \times 2 + 27 \div 9 \quad \rightarrow$ brackets

$= 13 \times 2 + 27 \div 9 \quad \rightarrow$ multiply

$= 26 + 3 \qquad\qquad\quad$ and divide

$= 29 \qquad\qquad\qquad \rightarrow$ add

4 $1 - (4 \cdot 8 - 0 \cdot 3) \times 6$

$1 - (4 \cdot 8 - 0 \cdot 3) \times 6 \quad \rightarrow$ brackets

$= 1 - 4 \cdot 5 \times 6 \qquad\quad \rightarrow$ multiply

$= 1 - 27 \qquad\qquad\quad \rightarrow$ subtract

$= -26$

5 $\dfrac{4 - 10}{-2 - (-5)}$

$\dfrac{4 - 10}{-2 - (-5)} \quad \rightarrow$ simplify numerator and denominator

$= \dfrac{-6}{-2 + 5}$

$= \dfrac{-6}{3} \quad \rightarrow$ divide

$= -2$

6 $(2 + 3 \times 6) \div (-1 - \dfrac{1}{2}$ of $8)$

$(2 + 3 \times 6) \div (-1 - \dfrac{1}{2}$ of $8) \quad \rightarrow$ BODMAS to simplify each bracket

$= (2 + 18) \div (-1 - 4)$

$= 20 \div (-5) \qquad\quad \rightarrow$ divide

$= -4$

Exercise 1N

1 Calculate

a) $2 + 4 \times 7$

b) $25 - \frac{1}{3}$ of 21

c) $3 \times (2 + 4)$

d) $15 \div (17 - 2)$

e) $12 - 5 \times 2$

f) $(9 + 7) \div 8$

g) $11 - 11 \div 11$

h) $\frac{1}{3}$ of $6 + 9$

i) $(2 + 9) \times (7 - 1)$

j) $6 \times 8 - 4 \times 7$

k) $8 + 3 \times 5 - 1$

l) $15 + \frac{1}{4}$ of $12 + 8$

2 Use the table below to convert your answers into letters. The answer to each question should then spell a word.

An example is completed for you.

A	B	C	D	E	F	G	H	I	J	K	L	M	N	O	P	Q	R	S	T	U	V	W	X	Y	Z
1	2	3	4	5	6	7	8	9	10	11	12	13	14	15	16	17	18	19	20	21	22	23	24	25	26

Example:

$10 - (5 + 1)$
$= 10 - 6$
$= 4 \rightarrow D$

$2 + \frac{1}{2}$ of 26
$= 2 + 13$
$= 15 \rightarrow O$

$15 - 4 \times 2$
$= 15 - 8$
$= 7 \rightarrow G$

The answer is DOG.

a) $9 + 2 \times 8$ $3 \times (7 - 2)$ $(15 + 9) \div (10 - 8)$ $22 - \frac{1}{6}$ of $30 - 6$

b) $24 - (13 - 8)$ $\frac{1}{3}$ of $24 + 12$ $8 - (5 + 2)$ $(4 + 2) \times (7 - 4)$ $9 + \frac{1}{3}$ of $15 - 3$

c) $10 - 4 \div 2$ $1 + 5 \times 2 - 6$ $2 \times 7 - (9 + 4)$ $20 - 2 \times 3 + \frac{1}{2}$ of 8 $30 - 3 \times 6 + 4 \times 2$

d) Write a question which uses BODMAS and simplifies to give your name.

3 Calculate:

a) $3 + 1 \cdot 5 \times 10$

b) $(6 \cdot 8 + 5 \cdot 6) \div 4$

c) $10 \div (-2) - 3$

d) $7 - (9 + 3)$

e) $\dfrac{15 + 2 \times 6}{3}$

f) $\dfrac{72}{30 - 3 \times 7}$

g) $\dfrac{17 + 20 \div 4}{2 \times (7 + 4)}$

h) $\dfrac{230 - 2 \times 15}{4 + \frac{1}{5} \text{ of } 30}$

4 Evaluate:

a) $(1 + 8 \times 3) \div 5$

b) $4 \times (9 - 3 \times 2)$

c) $(28 - 9 \times 2) - (3 - (2 - 1))$

d) $(4 + 9 \times 2) \times (10 + \frac{1}{2}$ of $16 - 8)$

e) $(5 \times 7 - 3) \div (18 - 7 \times 2)$

f) $\dfrac{3 - (3 + 5 \times 6)}{1 + 2 \times 3 - 2}$

Check-up

1 Find:

    a) 15·9 + 8·41   b) 2·15 − 0·917   c) 12·8 × 3   d) 2·691 × 8

    e) 45·6 × 6   f) 40·6 ÷ 7   g) 22 ÷ 8   h) 8·1 ÷ 5

2 Mohammad bought three tins of paint. Two cost £8·70 each and the other cost £8·95.

    a) Estimate whether Mohammad can buy these for £25.

    b) Calculate how much change Mohammad will get from £30.

3 The famous mathematical number π can be approximated by $\frac{22}{7}$ = 3·14285714 .... Round $\frac{22}{7}$ to

    a) 2 decimal places   b) 3 decimal places.

4 Use the information in the table to calculate, in litres,

    a) the volume of one teaspoon and one tablespoon

    b) the difference between one pint and one tablespoon

    c) the total volume of one teaspoon, one fluid ounce and one pint

Measure	Volume in litres
One teaspoon	0·006
One fluid ounce	0·028
One tablespoon	0·178
One pint	0·568

    d) how much water will be left if a one litre bottle of water is used to fill one of each of the measures.

5 Calculate:

    a) 13·9 × 10   b) 423·618 × 100   c) 15·64718 × 1000   d) 1290 ÷ 10

    e) 635 ÷ 100   f) 182·1 ÷ 1000   g) 2·83 × 70   h) 9·108 × 400

    i) 16·71 × 8000   j) 693 ÷ 30   k) 4512 ÷ 500   l) 94 ÷ 4000

6 At an awards ceremony, 300 certificates are issued. It costs £0·19 to print each certificate. How much does it cost to print all the certificates?

7 Malcolm is a panel-beater. He bought 400 pens inscribed with his business details to give to his customers. The total cost of the pens was £304. How much did it cost Malcolm for

    a) 1 pen   b) 125 pens?

8 The length of a hotel swimming pool is 36 metres.
   How far did Kamal swim if he completed 45 lengths of the pool?

9 Eilidh has 23 cousins. She wants to buy them all a calendar with the date of her birthday marked on it so that they do not forget when her birthday is. Each calendar costs £2·79.

   How much will it cost Eilidh to buy a calendar for each of her cousins?

10 On average, Hannah sends 29 messages a day to her friends. If Hannah continues at this rate for 365 days

    a) estimate how many messages Hannah will send

    b) calculate how many messages she will send.

   Hannah is now on a call plan where she gets unlimited texts, however, before this she was charged £0·05 for every message she sent.

    c) How much would Hannah have to pay for 365 days on her old rate?

11 Find:

a) $6 - 10$ b) $-15 + 3$ c) $5 + (-2)$ d) $8 - (-6)$

e) $-7 + (-1)$ f) $-4 - (-5)$ g) $-16 + (-7)$ h) $-20 - (-11)$

12 Mekwi is in Southend and notices the temperature is 8°C. He calls his friend in Canada who says that it is −34°C there! What is the difference in temperature between Southend and Canada?

13 Calculate:

a) $7 \times (-2)$ b) $5 \times (-4)$ c) -6×8 d) -1×9

e) $-110 \div (-11)$ f) $24 \div (-6)$ g) $-35 \div 5$ h) $-160 \div (-8)$

14 Simplify:

a) $7 + 5 \times 10$ b) $(6 + 8) \div (9 - 2)$ c) $1 + \frac{1}{4}$ of $12 + 8$ d) $\dfrac{21 - 15 \div 3}{3 + 2 \times 2 + 1}$

15 Mairi, Thomas and Fiona have these cards:

Mairi's card		
16	−3	2
5	−4	−1
−5	10	−12

Thomas' card		
−5	−4	−1
−3	16	6
−9	2	10

Fiona's card		
−5	2	10
−4	−3	7
−11	−1	16

Graeme reads out the calculations in the order shown in the table.

If Mairi, Thomas or Fiona have the answer to Graeme's question on their card then they circle it.

Copy the cards into your jotter and circle the correct answers as you work down the questions.

Who is first to complete a:

a) horizontal line

b) vertical line

c) full card?

Questions
$10 + (-3)$
$14 - 2 \times 9 + 1$
$-5 + 11$
$-10 \div 2$
$2 + 3 \times 6 - 4$
$-30 + \frac{1}{2}$ of $50 + 10$
$2 \times 4 - 1 \times (-2)$
$5 + 3 \times (-1)$
$(-1 + 3) \times (-3 + 1)$
$-15 + \frac{1}{3}$ of $(7 + 5)$
$(1 + 2 \times 4) \div (3 - 6 \times 2)$
$-3 \times (5 - 2)$
$2 \times (-3) \times 2$

16 You have

2   6   1   7   +   ( )   ×   −

Digits cannot be written side by side to make 26, 261, 71, etc.

You must use each number and **operator** once and only once.

As an example, a target of 47 can be obtained from the calculation $(1 + 6) \times 7 - 2$.

Find a calculation which gives

a) 17   b) −27

c) the highest possible number   d) the lowest possible number.

2 Multiples, factors, powers and roots

▶ Lowest common multiple

Multiples of a number are the numbers that can be divided by it without leaving a remainder (i.e. the 'stations' in that number's times table). To find the **lowest common multiple** (LCM) of two numbers:

- List the first few multiples of each number.
- The first number to appear in both lists is the lowest common multiple.

Worked examples

1 List the first five multiples of 6.

The first five multiples of 6 are 6, 12, 18, 24 and 30.

2 List the first five multiples of 17.

The first five multiples of 17 are 17, 34, 51, 68 and 85.

3 Is 155 a multiple of 3?

$$3\overline{)1^15^25}$$ 51r2

Since 3 does not divide exactly into 155, 155 is not a multiple of 3.

4 What is the lowest common multiple of 6 and 8?

Start by listing the first few multiples of each number:

multiples of 6: 6, 12, 18, ⑳, 30...

multiples of 8: 8, 16, ⑳, 32, 40...

The lowest common multiple of 6 and 8 is 24.

5 What is the lowest common multiple of 3, 5 and 12?

Multiples of 12: 12, 24, 36, 48, ⑥⓪, 72...

Multiples of 5: 5, 10, 15, 20, 25, 30, 35, 40, 45, 50, 55, ⑥⓪

Notice that 60 is the lowest common multiple of 5 and 12. 60 is also a multiple of 3, so the lowest common multiple of 3, 5 and 12 is 60.

Exercise 2A

1 List the first five multiples of each number.

a) 7   b) 3   c) 9   d) 8   e) 12

f) 15   g) 40   h) 18   i) 26   j) 104

2 What is the 15th multiple of 7?

3 What is the 52nd multiple of 39?

4 Is 237 a multiple of 3?

5 Is 260 a multiple of 6?

6 For each number, state whether it is a multiple of 2, 5, both or neither.

a) 24   b) 75   c) 40   d) 19   e) 400   f) 1965   ➜

7   Explain how you can quickly identify whether a given number is a multiple of 2.

8   Explain how you can quickly identify whether a given number is a multiple of 5.

9   A quick way of checking whether a number is a multiple of 3 is to sum its digits. If the **digit sum** is a multiple of 3 then so is the original number. For example, to check whether 798 is a multiple of 3, check 7 + 9 + 8 = 24. Since 24 is a multiple of 3, so is 798. **Note: this method only works for multiples of 3 or 9!**

   Using the same method, check whether these numbers are multiples of 3.

   a)   237   b)   402   c)   566   d)   1002
   e)   3551   f)   19 287   g)   131 313   h)   6851

10 Find the lowest common multiple of the numbers in each pair.

   a)   2 and 9   b)   4 and 6   c)   5 and 7   d)   6 and 8
   e)   5 and 20   f)   12 and 16   g)   8 and 14   h)   15 and 8
   i)   21 and 14   j)   13 and 3   k)   8 and 42   l)   11 and 13

11 Find the lowest common multiple of

   a)   4, 5 and 6   b)   2, 7 and 10   c)   3, 5 and 8   d)   7, 8 and 12

12 Three sets of lights flash at different rates. The first set flashes every 4 seconds, the second set every 6 seconds and the third set every 9 seconds. If all three lights have just flashed together, how long will it be until they next flash together?

13 Rory has forgotten his password. He knows it is a 4-digit number with the following properties:

   ●   It is a multiple of 3 and 5.

   ●   The first and last digits are the same.

   ●   The middle two digits are the lowest common multiple of 5 and 16.

   What is Rory's password?

14 Cats Winnie and Willow visit their food bowls at regular intervals. Willow visits his every 5 hours and Winnie visits hers every 8 hours. If they both visit their food bowl at 7am on Monday, what is the next time they will visit their bowls together, and on which day?

15 A bus station runs three different buses on a Saturday. The A1 sets off from the bus station every 45 minutes. The A2 sets off every 18 minutes. The B6 sets off every 20 minutes. If all three buses have just left the station at 10.30 am, what is the next time that they will all leave together?

16 A teacher is making up packed lunches for a school trip. Each packed lunch contains an apple, a bottle of water and a bag of crisps. Crisps are sold in multipacks of 12, apples are sold in bags of 5 and bottled water comes in packs of 8. What is the smallest number of packed lunches that could be made up with no food or water left over?

17 Priyanka wants to organise her pupils' posters on the classroom wall. If she puts them up in rows of 5 there are two posters left over. If she puts them up in rows of 6, there are 3 posters left over. If she puts them up in rows of 9, they fit perfectly. If there are fewer than 30 pupils in Priyanka's class, how many posters does she have?

18 Jenny has some treats for her horse, Kodiak. If she arranges them in rows of 5, there are 2 left over. If she arranges them in rows of 9, there are 3 left over. If she arranges them in rows of 13, there are 5 left over. What is the smallest number of treats that Jenny could have?

▶ Highest common factor

Factors of a number are the values that can divide into it without leaving a remainder. When finding the factors of a number it is often useful to think of them in pairs that multiply together to give that number. To find the **highest common factor** (HCF) of two numbers:

- List the factors of each.
- Identify the largest number that appears in both lists.

Worked examples

1   Find all the factors of

a)   18

List the pairs of numbers that multiply to give 18:

$$
\begin{array}{ll}
\underline{\quad 18 \quad} \\
1 & 18 \\
2 & 9 \\
3 & 6 \\
\end{array}
$$

Now list them in order: The factors of 18 are 1, 2, 3, 6, 9 and 18.

b)   36.

List the pairs of numbers that multiply to give 36:

$$
\begin{array}{ll}
\underline{\quad 36 \quad} \\
1 & 36 \\
2 & 18 \\
3 & 12 \\
4 & 9 \\
6 & 6 \\
\end{array}
$$

The factors of 36 are 1, 2, 3, 4, 6, 9, 12, 18 and 36.

(Note that we do not list 6 twice.)

2   Is 7 a factor of 85?

$$
\begin{array}{r}
1\ 2\ \text{r}1 \\
7\overline{)8^15}
\end{array}
$$

Since 85 is not divisible by 7, 7 is not a factor of 85.

3   Find the highest common factor of

a)   12 and 21

List the factors of each number.

$$
\begin{array}{ll}
\underline{\quad 12 \quad} & \underline{\quad 21 \quad} \\
1 \quad 12 & 1 \quad 21 \\
2 \quad 6 & ③ \quad 7 \\
③ \quad 4 &
\end{array}
$$

Look for the largest number that appears in both lists.

The highest common factor of 12 and 21 is 3.

We write HCF(12, 21) = 3.

b)   14, 28 and 42.

List the factors of each number.

$$
\begin{array}{lll}
\underline{\ 14\ } & \underline{\ 28\ } & \underline{\ 42\ } \\
1 \ ⑭ & 1 \ 28 & 1 \ 42 \\
2 \ 7 & 2 \ ⑭ & 2 \ 21 \\
 & 4 \ 7 & 3 \ ⑭ \\
 & & 6 \ 7
\end{array}
$$

Look for the largest number that appears in all the lists.

The highest common factor of 14, 28 and 42 is 14.

HCF(14, 28, 42) = 14.

Exercise 2B

1   List all the factors of each number.

a) 10   b) 15   c) 24   d) 39

e) 17   f) 100   g) 81   h) 54

i) 180   j) 400   k) 105   l) 1

2   Some of the numbers in question 1 have an odd number of **distinct** (different) factors. What is the connection between them? Can you give three more examples of numbers that will have an odd number of distinct factors?

3   Find the highest common factor of the numbers in each pair.

a) 14 and 35   b) 8 and 20   c) 15 and 60   d) 24 and 56

e) 27 and 81   f) 23 and 30   g) 78 and 16   h) 13 and 52

i) 70 and 105   j) 51 and 85   k) 120 and 18   l) 56 and 133

4   Find the highest common factor of

a) 18, 30 and 45   b) 28, 35 and 49   c) 24, 96 and 180.

5   Copy and complete the table. Comment on any connections you notice.

1st number	2nd number	Product	Highest common factor	Lowest common multiple
6	15	90	3	30
8	12			
20	50			
18	36			
22	33			

6   The proper factors of a number are the factors other than the number itself.

A **perfect number** is equal to the sum of its proper factors. For example, the first perfect number is 6. The proper factors of 6 are 1, 2 and 3, and $1 + 2 + 3 = 6$.

Find the second perfect number. (**Hint:** it is a two-digit number.)

Note: Nobody has ever found an odd perfect number, and it is an unknown whether any exist!

7   A number is **abundant** if the sum of its proper factors is greater than the number itself. Show that 24 and 60 are abundant numbers. Can you think of a reason why this is useful for our system of measuring time (24 hours in a day, 60 minutes in an hour)?

8   Neil asks his teacher her birthday. She tells him "The day is the highest common factor of 60 and 105. The month is the highest common factor of 55 and 154." When is Neil's teacher's birthday?

9   Find a pair of numbers that have the following property: the product of their highest common factor and lowest common multiple is equal to the product of the numbers themselves.

Can you find three more such pairs?

Can you make a conjecture (a theory based on the information you have) about this property?

▶ Prime numbers

- A number is prime if it has exactly two factors: itself and 1.
- 1 is **not** a prime number as it has only one factor.
- Numbers with more than two factors are called composite numbers.

Worked examples

1   Determine whether the number 18 is prime.

Try to think of a factor of 18 other than itself and 1. The factors of 18 are 1, 2, 3, 6, 9 and 18.

Since 18 has more than two factors, it is **not** prime.

2   Determine whether the number 7 is prime.

The only factors of 7 are 1 and 7, therefore 7 **is** prime.

Exercise 2C

1   Determine whether each number is prime.

a) 5   b) 6   c) 15   d) 2   e) 11   f) 21

g) 23   h) 40   i) 49   j) 121   k) 159   l) 477

2   The Sieve of Eratosthenes is a method for finding prime numbers. Follow these steps to use it.

1	2	3	4	5	6	7	8	9	10
11	12	13	14	15	16	17	18	19	20
21	22	23	24	25	26	27	28	29	30
31	32	33	34	35	36	37	38	39	40
41	42	43	44	45	46	47	48	49	50
51	52	53	54	55	56	57	58	59	60
61	62	63	64	65	66	67	68	69	70
71	72	73	74	75	76	77	78	79	80
81	82	83	84	85	86	87	88	89	90
91	92	93	94	95	96	97	98	99	100

- Copy the 100 square into your jotter.
- Neatly score out the number 1 (as it isn't prime).
- Circle the next number along (2). This is our first prime number.
- Now work your way through the square, neatly scoring out every number that is a multiple of 2 (as they can't be prime if 2 is a factor).
- The smallest number that hasn't been scored out is our next prime number. Circle it and then score through every number that is a multiple of it.
- Repeat this process until every number in the square has been circled or scored out. You should now have found the first 25 prime numbers!

3   How many prime numbers are there:

a) between 30 and 40   b) between 80 and 90   c) less than 20?

4   Twin primes are a pair of prime numbers that differ by 2, for example 3 and 5, or 5 and 7.

a) Find six more twin primes less than 100.   b) Find two twin primes larger than 100.

5   Find two prime numbers that multiply together to give each number.

a) 10   b) 21   c) 55   d) 34   e) 77   f) 143   ➡

6   The 'Goldbach Conjecture' is an unsolved problem in mathematics. It says that every even integer larger than 2 can be written as the sum of two prime numbers. For example, $16 = 5 + 11$ and $30 = 13 + 17$. Find a way to write each of these numbers as the sum of two prime numbers.

   a) 8   b) 20   c) 28   d) 40   e) 64   f) 100

7   Steve notices that when adding $4^2 + 5^2$ he gets the answer 41, a prime number. He tries adding more combinations of odd and even square numbers and gets a prime each time. His friend Sachin thinks this will not always be the case. Can you find an odd square number and an even square number that **do not** add up to a prime number?

8   Use your calculator to find how many ways you can arrange the digits 1, 2 and 3 to give a three-digit prime number.

9   Find all the two-digit prime numbers that are still prime when their digits are read in reverse order. (**Hint:** there are nine of them.)

10 The digit sum of a number is the number obtained by adding its individual digits. For example, the digit sum of 235 is $2 + 3 + 5 = 10$. Find all ten two-digit prime numbers whose digit sum is also prime.

11 Copy each grid then follow the instructions underneath.

a)

20	3	14	8
12	32	100	11
35	70	19	71
29	18	6	4

b)

55	4	17	30
24	59	110	97
22	5	16	12
91	15	37	121

Score out: • prime numbers
             • multiples of 5
             • factors of 24.

What is the sum of the remaining numbers?

Score out: • prime numbers
             • factors of 60
             • multiples of 11.

What is the sum of the remaining numbers?

Two or more positive integers are coprime (or relatively prime) if their highest common factor is 1. Note that the integers themselves need not be prime for this condition to be met. Being able to spot whether two numbers are coprime is particularly useful when simplifying fractions.

Worked examples

1   Determine whether the numbers 15 and 28 are coprime.

   The factors of 15 are 1, 3, 5 and 15.

   The factors of 28 are 1, 2, 4, 7, 14 and 28.

   Since their highest common factor is 1, 15 and 28 **are** coprime.

2   Determine whether the numbers 9 and 24 are coprime.

   The factors of 9 are 1, 3 and 9.

   The factors of 24 are 1, 2, 3, 4, 6, 8, 12 and 24.

   Since their highest common factor is 3, 9 and 24 are **not** coprime.

Exercise 2D

1   Determine whether the numbers in each pair are coprime.

   a) 10 and 21   b) 24 and 33   c) 48 and 9   d) 54 and 135   e) 77 and 450

▶ Powers and roots

The **index**, or **power**, of a number tells you how many 'lots' of the number have to be multiplied together.

For example, the number 2^4 has a **base** of 2 and a power of 4, and means $2 \times 2 \times 2 \times 2$ (i.e. four lots of 2 all multiplied together). Say this number as 'two to the power of four'.

A number raised to the power 2 is **squared** and a number raised to the power 3 is **cubed**. For example, read 5^2 as 'five squared' and 6^3 as 'six cubed'.

In this chapter, you will only consider powers that are positive integers.

Worked examples

1 Write in index form:

a) $7 \times 7 \times 7$

$= 7^3$

b) $3 \times 3 \times 3 \times 3 \times 3$

$= 3^5$

c) 17×17

$= 17^2$

d) $a \times a \times a$

$= a^3$

2 Write in index form:

a) $2 \times 2 \times 2 \times 5 \times 5$

$= 2^3 \times 5^2$

b) $3 \times 3 \times 7 \times 7 \times 11$

$= 3^2 \times 7^2 \times 11$

3 Evaluate:

a) 5^2

$= 5 \times 5$

$= 25$

b) 2^4

$= 2 \times 2 \times 2 \times 2$

$= 16$

c) $(-7)^2$

$= (-7) \times (-7)$

$= 49$

d) $4^3 - 3^2$

$= 64 - 9$

$= 55$

Exercise 2E

1 Write in index form:

a) 9×9

b) $8 \times 8 \times 8$

c) $5 \times 5 \times 5 \times 5$

d) $2 \times 2 \times 2 \times 2$

e) $3 \times 3 \times 3$

f) $11 \times 11 \times 11 \times 11$

g) $a \times a \times a \times a$

h) $b \times b \times b$

2 Write these in index form.

a) $2 \times 2 \times 2 \times 2 \times 3 \times 3$

b) $5 \times 5 \times 7 \times 7 \times 13 \times 13$

c) $2 \times 3 \times 3 \times 7 \times 7 \times 7$

d) $11 \times 11 \times 17 \times 17 \times 17$

e) $23 \times 23 \times 29 \times 29 \times 29$

f) $2 \times 2 \times 2 \times 2 \times 3 \times 3 \times 5 \times 7 \times 7$

3 Evaluate:

a) 2^3

b) 6^2

c) 3^4

d) 4^3

e) 9^2

f) 1^8

g) 6^3

h) 2^5

i) 10^6

j) 7^3

k) 4^4

l) 20^3

4 Evaluate:

a) $5^2 + 3^2$

b) $4^2 - 2^3$

c) $10^3 + 3^3$

d) $8^2 - 3^3$

e) $2^4 + 2^3 + 2^2$

f) $6^2 - 11$

g) $(3 \times 5 - 9)^2$

h) $7 + 2^3 \times 3$

i) $(5 - 2)^3 + 8^2$

j) 5×3^2

k) $5 \times 2^2 - 3$

l) 3×2^4

➜

5   Evaluate these powers of negative numbers.

   a) $(-8)^2$ b) $(-3)^3$ c) $(-10)^4$ d) $(-6)^3$ e) $(-1)^7$ f) $(-2)^5$ g) $(-20)^2$

6   Can you spot a connection between the powers in question 5 and the sign of the answers you achieved? Can you explain why this will always be the case for powers of negative numbers?

7   A **Mersenne number** is one less than a power of two, for example $2^3 - 1$. Evaluate these Mersenne numbers and state whether they are prime.

   a) $2^4 - 1$ b) $2^5 - 1$ c) $2^7 - 1$ d) $2^8 - 1$

8   Can a power of an odd number ever be even? Explain your answer.

When finding large powers of numbers, it is often necessary to use a calculator. On a standard scientific calculator there is a button with the symbol x^y or $x^\square$. This allows you to find powers of larger numbers.

9   Use your calculator to evaluate these powers.

   a) 12^7 b) 4^9 c) 18^5 d) 31^2 e) 17^3 f) 25^4

   g) 2^{10} h) 46^4 i) 16^5 j) 7^6 k) 14^3 l) 9^5

10   Use your calculator to evaluate these expressions.

   a) $6^4 - 3^3$ b) $12^2 + 5^4$ c) $18^4 - 17^4$ d) $5 \times 7^3 - 8$

   e) $2^3 + 3^3 + 4^3$ f) $5^2 \times 6^3$ g) $15^4 \div 5^3$ h) $12^6 \div 2^{12}$

>> Roots

Worked examples

1   Which numbers give the answer 81 when squared?

   The square roots of a number are the values which give that number when squared.

   The positive square root of 81 is 9. We can write $\sqrt{81} = 9$. But −9 will also give 81 when squared. The square roots of 81 are 9 and −9.

2   Which number when cubed gives the answer 64?

   Remember that 'cubing' a number means multiplying it by itself, then by itself again.

   $4^3 = 64$, so the required number is 4. You can write $\sqrt[3]{64} = 4$.

   (Note that $(-4)^3 = (-4) \times (-4) \times (-4) = -64$, so −4 is not a cube root of 64.)

Exercise 2F

1   Which numbers, when squared, give these answers?

   a) 36   b) 4   c) 49   d) 100   e) 64

   f) 121   g) 400   h) 144   i) 225   j) 196

2   Which numbers, when cubed, give these answers?

   a) 8   b) 1   c) 1000   d) −27   e) 125

▶ Fundamental theorem of arithmetic

The **fundamental theorem of arithmetic** says that every positive integer larger than one is either prime or can be written as a product of prime numbers in a unique way. This is the **prime factorisation** of the number.

To write the prime factorisation of a given integer it can be helpful to use a **factor tree**.

- Write the original integer at the top, then split it into two 'branches' by finding a pair of factors of the integer.

- Continue to split each factor, stopping a branch every time you arrive at a prime number.

- Rewrite your factors in ascending order using index notation.

Worked examples

1   Find the prime factorisation of 18.

Write the original integer.

2   9   Split it into a pair of factors.

3   3   Continue until all branches end in prime numbers.

You can now write the prime factorisation as $18 = 2 \times 3 \times 3$

Finally, write any repeated primes in index form. $18 = 2 \times 3^2$

2   Find the prime factorisation of 840

$840 = 2 \times 2 \times 3 \times 7 \times 2 \times 5$

$\quad = 2 \times 2 \times 2 \times 3 \times 5 \times 7$

$\quad = 2^3 \times 3 \times 5 \times 7$

Notice that you will arrive at the same result even if you choose different factors of 840.

$840 = 2 \times 2 \times 2 \times 5 \times 3 \times 7$

$\quad = 2^3 \times 3 \times 5 \times 7$

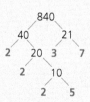

Exercise 2G

1   Find the prime factorisation of each integer.

a) 20   b) 45   c) 30   d) 63

e) 54   f) 84   g) 375   h) 40

i) 144   j) 132   k) 100   l) 231

m) 300   n) 588   o) 192   p) 700

2   For each pair of integers, find their prime factorisation and hence state whether they are coprime.

a) 80 and 207   b) 544 and 675   c) 744 and 945   d) 183 and 3040

You can use the prime factorisations of integers to find their lowest common multiples and highest common factors. Once you have the prime factors of both numbers you can arrange them in a Venn diagram.

Worked example

Use prime factorisation to find the lowest common multiple and highest common factor of 490 and 532.

First, use factor trees to find the prime factorisation of each integer.

$$490 = 2 \times 5 \times 7 \times 7$$
$$= 2 \times 5 \times 7^2$$

$$532 = 2 \times 2 \times 7 \times 19$$
$$= 2^2 \times 7 \times 19$$

Now arrange the factors of both integers in a Venn diagram.

Place any shared factors in the overlapping region.

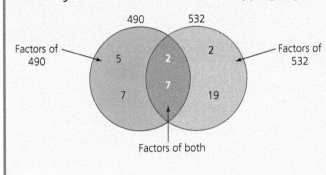

To find the highest common factor, multiply together the factors in the overlapping region.

The highest common factor of 490 and 532 is $2 \times 7 = 14$.

To find the lowest common multiple, multiply together the factors in every region.

The lowest common multiple of 490 and 532 is $5 \times 7 \times 2 \times 7 \times 2 \times 19 = 18620$.

Exercise 2H

You may use a calculator for this exercise.

1   For each pair of integers, use prime factorisation to find the highest common factor and lowest common multiple.

   a) 48 and 90   b) 30 and 38   c) 24 and 60   d) 300 and 81

   e) 729 and 36   f) 189 and 525   g) 484 and 200   h) 324 and 450

2   Two numbers have a highest common factor of 8 and a lowest common multiple of 48. Given that one of the numbers is 16, find the other.

3   Two numbers have a highest common factor of 3 and a lowest common multiple of 84. Given that one of the numbers is 21, find the other.

4   Two numbers have a highest common factor of 9 and a lowest common multiple of 126. Given that one of the numbers is 63, find the other.

Check-up 👍

1 List the first five multiples of each number.

   a) 4   b) 5   c) 11   d) 13

   e) 25   f) 60   g) 120   h) 115

2 Find the lowest common multiple of the numbers in each pair.

   a) 6 and 14   b) 12 and 10   c) 20 and 35   d) 27 and 63

3 Find the lowest common multiple of 4, 5 and 18.

4 Find the lowest common multiple of 6, 7 and 9.

5 Three remote control cars are doing laps of a circuit. The fastest takes 12 minutes to complete a lap, the next fastest takes 15 minutes and the slowest takes 20 minutes.

   a) If they all start at the same time, how long will it be until they all complete a lap at the same time?

   b) How many laps will each car have completed?

6 A mathematical musician is programming a drum track for his latest composition based on prime numbers. The bass drum sounds every third beat. The snare drum sounds every seventh beat. The triangle sounds every thirteenth beat.

   a) If the drums sound together on the first beat of the composition, on which beat do they next sound together?

   b) There are five beats to a bar. How many complete bars have there been by the second time the drums sound together?

7 Is 2937 a multiple of 3? Give a reason for your answer.

8 Is 1531 a multiple of 3? Give a reason for your answer.

9 Is 1004 a multiple of 5? Give a reason for your answer.

10 List all the factors of each number.

   a) 14   b) 16   c) 85   d) 220

11 Find the highest common factor of the numbers in each pair.

   a) 30 and 24   b) 140 and 56   c) 36 and 108   d) 75 and 300

12 Find the highest common factor of 40, 56 and 120.

13 Find the highest common factor of 21, 63 and 84.

14 State whether each of these numbers is prime. If it is not, list its factors.

   a) 41   b) 51   c) 26   d) 93

15 State whether the numbers in each pair are coprime. If they are not, write their highest common factor.

   a) 28 and 17   b) 32 and 24   c) 45 and 18   d) 77 and 121

16 Lisa is trying to remember the 4-digit combination for her safe where she keeps her favourite maths books. She remembers that the first digit is the highest common factor of 32 and 40. The second digit is the highest common factor of 24 and 18. The last two digits are the lowest common multiple of the first two digits. Can you find the combination for Lisa? ➔

17 Write these numbers in index form.

a) $6 \times 6 \times 6 \times 6$ b) $8 \times 8 \times 8$ c) 13×13 d) $2 \times 2 \times 2 \times 2 \times 2 \times 2 \times 2$

e) $4 \times 4 \times 4 \times 4 \times 4$ f) $53 \times 53 \times 53 \times 53$ g) $3 \times 3 \times 3$ h) $a \times a \times a \times a \times a$

18 Evaluate these powers.

a) 9^2 b) 3^3 c) 2^6 d) 5^4

e) 10^4 f) 7^3 g) $(-2)^4$ h) $(-1)^5$

19 Which positive integer gives 144 when squared?

20 Which positive integer gives 64 when cubed?

21 Find the prime factorisation of each number.

a) 50   b) 135   c) 81   d) 98

e) 378   f) 198   g) 264   h) 980

22 Use prime factorisation to find the highest common factor and lowest common multiple of each pair of integers.

a) 50 and 65   b) 18 and 33   c) 24 and 36   d) 81 and 36

e) 130 and 70   f) 23 and 30   g) 39 and 65   h) 200 and 75

23 Two numbers have a highest common factor of 12 and a lowest common multiple of 72. Given that one of the numbers is 24, find the other.

24 Copy and complete the grids below.

a)

15	18	7	31	14
9	11	21	16	12
45	23	42	30	8
6	32	20	54	2
75	90	1	4	17

b)

30	75	81	72	3
96	25	2	10	100
18	200	27	20	1
19	4	6	15	17
5	9	243	8	90

Score out:
- prime numbers
- factors of 60
- multiples of 5
- powers of 2.

What is the sum of the remaining numbers?

Score out:
- powers of 3
- factors of 75
- multiples of 6
- factors of 100.

What is the sum of the remaining numbers?

c)

98	2	92	44
6	7	51	32
16	22	5	76
83	14	101	13

d)

30	81	84	15
19	35	59	28
21	31	20	57
54	6	49	24

Score out:
- prime numbers
- numbers which are prime when their digits are reversed.

What is the sum of the remaining numbers?

Score out:
- prime numbers
- factors of 120
- multiples of 7.

What is the sum of the remaining numbers?

3 Fractions, decimals and percentages

▶ Simplifying fractions

Equivalent fractions have the same value but are written differently. For example, $\dfrac{50}{100}$, $\dfrac{10}{20}$ and $\dfrac{5}{10}$ are all equivalent as they are all equal to $\dfrac{1}{2}$. Always simplify fractions as far as possible.

- Identify the highest common factor of the numerator and denominator.
- Divide both the numerator and denominator by the highest common factor.

> **Worked examples**
>
> 1. $\dfrac{6}{38}$ HCF = 2 so divide by 2
>
> $\dfrac{\cancel{6}^{3}}{\cancel{38}_{19}}$
>
> $= \dfrac{3}{19}$
>
> 2. $\dfrac{14}{7}$ HCF = 7 so divide by 7
>
> $\dfrac{\cancel{14}^{2}}{\cancel{7}_{1}}$
>
> $= \dfrac{2}{1}$
>
> $= 2$
>
>   You cannot leave 1 on the bottom of a fraction as this is unsimplified. Any number divided by 1 is just itself.
>
> 3. $\dfrac{90}{36}$ HCF = 18 so divide by 18
>
> $\dfrac{\cancel{90}^{5}}{\cancel{36}_{2}}$
>
> $= \dfrac{5}{2}$
>
>   If you did not spot 18 as the HCF you can simplify in stages.
>
> $\dfrac{\cancel{90}^{45}}{{}_{18}\cancel{36}}$ divide by 2
>
> $= \dfrac{\cancel{45}^{5}}{{}_{2}\cancel{18}}$ divide by 9
>
> $= \dfrac{5}{2}$

Exercise 3A

1. Copy and complete.

   a) $\dfrac{15}{25}$ (HCF = 5)

   $= \dfrac{3}{\square}$

   b) $\dfrac{24}{39}$ (HCF = 3)

   $= \dfrac{\square}{13}$

   c) $\dfrac{49}{84}$ (HCF = 7)

   $= \dfrac{7}{\square}$

   d) $\dfrac{90}{30}$ (HCF = 30)

   $= \square$

2. Simplify:

   a) $\dfrac{27}{9}$ b) $\dfrac{8}{12}$ c) $\dfrac{10}{200}$ d) $\dfrac{5}{60}$ e) $\dfrac{40}{36}$ f) $\dfrac{56}{49}$ g) $\dfrac{24}{72}$

3. Arizona has run 350 m in a 1500 m race. What fraction of the race has she completed?

4. From a class of 50 pupils, 44 pupils completed their Gold Duke of Edinburgh Award. What fraction of the class successfully completed the award?

5. Medicine is given in 5 ml doses. What fraction of a 100 ml bottle has been given after 4 doses?

6. In a company, 27 employees work part-time. If there is a total workforce of 72 people, what fraction of employees work full-time?

7. What fraction of 5 hours is 270 minutes? (Be careful with the units!)

▶ Decimals to fractions and percentages

The **decimal counting system** is based on place value.

- To write a decimal as a fraction, use the place value of the last digit to determine the denominator of the fraction. Simplify where possible.

Units	.	Tenths	Hundredths	Thousandths
0	.	$\dfrac{1}{10}$	$\dfrac{1}{100}$	$\dfrac{1}{1000}$

- To convert a decimal to a percentage, write as a fraction and then find an equivalent fraction with 100 as the denominator. The numerator is the percentage.

Worked examples 1 Write as fractions:

a) 0·6
$= \dfrac{6}{10}$
$= \dfrac{3}{5}$

The first decimal place is tenths so the denominator is 10.

b) 0·13
$= \dfrac{13}{100}$

The second decimal place is hundredths so the denominator is 100.

c) 0·425
$= \dfrac{425^{17}}{{}_{40}\cancel{1000}}$
$= \dfrac{17}{40}$

The third decimal place is thousandths so the denominator is 1000.

2 Convert these decimals to percentages.

a) 0.4
$= \dfrac{4^{\times 10}}{10_{\times 10}}$
$= \dfrac{40}{100}$
$= 40\%$

b) 0.78
$= \dfrac{78}{100}$
$= 78\%$

c) 0.029
$= \dfrac{29^{\div 10}}{1000_{\div 10}}$
$= \dfrac{2·9}{100}$
$= 2·9\%$

d) 0.008
$= \dfrac{8^{\div 10}}{1000_{\div 10}}$
$= \dfrac{0·8}{100}$
$= 0·8\%$

Note that in each of these examples, the working is equivalent to multiplying the decimal by 100%.

Exercise 3B

1 Convert to fractions in their simplest form.

   a) 0·1   b) 0·2   c) 0·8   d) 0·56   e) 0·75   f) 0·11

   g) 0·99   h) 0·003   i) 0·005   j) 0·016   k) 0·048   l) 0·102

2 Complete each decimal calculation and then match the answer to its simplified fraction.

0·2 + 0·1	0·8 − 0·32	0·96 ÷ 8	0·24 × 3	0·02 × 3	0·209 + 0·159	0·9 + 0·02	1 − 0·275
$\dfrac{3}{50}$	$\dfrac{46}{125}$	$\dfrac{12}{25}$	$\dfrac{29}{40}$	$\dfrac{3}{10}$	$\dfrac{23}{25}$	$\dfrac{18}{25}$	$\dfrac{3}{25}$

3 Convert to a percentage.

   a) 0·5   b) 0·9   c) 0·02   d) 0·12   e) 0·64   f) 0·004

   g) 0·045   h) 0·67 − 0·01   i) 0·04 × 2   j) 0·7 + 0·106   k) 0·82 ÷ 4   l) 0·318 × 3

▶ Percentages to fractions and decimals

- To convert a percentage to a fraction, write it as a fraction out of 100 and simplify (if possible).
- To write a percentage as a decimal, divide by 100.

Worked examples

1 Write 39% and 62% as fractions.

$$39\% = \frac{39}{100} \qquad 62\% = \frac{\cancel{62}^{31}}{\cancel{100}_{50}}$$

$$= \frac{31}{50}$$

2 Write 44% and 12.5% as decimals.

$$44\% = \frac{44}{100} \qquad 12 \cdot 5\% = \frac{12 \cdot 5}{100}$$

$$= 44 \div 100 \qquad = 12 \cdot 5 \div 100$$

$$= 0 \cdot 44 \qquad = 0 \cdot 125$$

Exercise 3C

1 Write each percentage as a fraction.

   a) 9%   b) 19%   c) 51%   d) 67%   e) 91%   f) 17%   g) 37%

   h) 8%   i) 46%   j) 82%   k) 95%   l) 68%   m) 36%   n) 54%

2 Write each percentage as a decimal.

   a) 6%   b) 30%   c) 72%   d) 59%   e) 83%   f) 92%   g) 16%

   h) 10·5%   i) 76·2%   j) 94·38%   k) 44·31%   l) 7·16%   m) 0·8%   n) 0·05%

3 Raman completed 34% of his workout before taking a break. What fraction of his workout did he complete?

4 In a restaurant, 62% of diners ordered the daily special. What fraction did not have the daily special?

5 In a survey, 71% of pupils answered that they walk to school, 14% cycle and the remainder use public transport. What fraction of pupils use public transport?

6 Copy this grid.

   a) There are four places when numbers in a row (horizontally, vertically or diagonally) add up to 1.

      One is already shaded for you. Find and shade the other three.

   b) What is the smallest total that can be made from any five squares? Write your answer as a percentage.

   c) Find four adjoining numbers that add to 1·81.

0·4	71%	14%	0·3	19%	24%	0·57
0·37	12%	62%	59%	0·46	61%	2%
33%	9%	0·38	0·17	16%	0·34	91%
0·02	3%	0·67	10%	20%	0·15	0·77
54%	0·87	13%	0·65	0·14	53%	12%
0·83	67%	0·55	19%	34%	8%	0·01

▶ Fractions to decimals and percentages

- To convert a fraction to a decimal, divide the numerator by the denominator.
- To write a fraction as a percentage the method depends upon the fraction given. Either:
  - find an equivalent fraction with 100 on the denominator or
  - convert to a decimal before multiplying by 100%.

Worked examples

1 Write $\dfrac{7}{8}$ and $\dfrac{17}{5}$ as decimals.

$$8\overline{)7.^70^60^40}\quad \text{giving } 0.875$$

$$5\overline{)17.^20}\quad \text{giving } 3.4$$

$\dfrac{7}{8} = 0.875 \qquad \dfrac{17}{5} = 3.4$

2 Find $\dfrac{4}{7}$ as a decimal to two decimal places.

$$7\overline{)4.^40^50^10^30}\quad \text{giving } 0.571\ldots$$

$\dfrac{4}{7} = 0.57$ to two decimal places

3 Convert $\dfrac{3}{10}$, $\dfrac{19}{20}$ and $\dfrac{3}{8}$ to percentages.

$\dfrac{3^{\times 10}}{10_{\times 10}}$ $\qquad$ $\dfrac{19^{\times 5}}{20_{\times 5}}$ $\qquad$ $8\overline{)3.^30^60^40}\quad \text{giving } 0.375$

$= \dfrac{30}{100}$ $\qquad = \dfrac{95}{100}$ $\qquad \dfrac{3}{8} = 0.375$

$= 30\%$ $\qquad = 95\%$ $\qquad 0.375 \times 100\%$

$\qquad\qquad\qquad\qquad\qquad\quad = 37.5\%$

Exercise 3D

1 Convert to decimals.

a) $\dfrac{9}{10}$ b) $\dfrac{3}{5}$ c) $\dfrac{7}{2}$ d) $\dfrac{5}{8}$ e) $\dfrac{11}{5}$ f) $\dfrac{9}{20}$ g) $\dfrac{8}{50}$

2 Write as a decimal, rounding your answers to three decimal places.

a) $\dfrac{2}{3}$ b) $\dfrac{5}{9}$ c) $\dfrac{6}{7}$ d) $\dfrac{8}{9}$ e) $\dfrac{10}{7}$ f) $\dfrac{14}{3}$ g) $\dfrac{19}{30}$

3 Convert to percentages.

a) $\dfrac{29}{50}$ b) $\dfrac{7}{10}$ c) $\dfrac{11}{20}$ d) $\dfrac{4}{5}$ e) $\dfrac{1}{8}$ f) $\dfrac{13}{40}$ g) $\dfrac{49}{80}$

4 By converting each number to the same form (either a fraction, decimal or percentage), write in order from smallest to largest:

a) 0.52, 58%, $\dfrac{51}{100}$, 0.57, $\dfrac{14}{25}$

b) 87%, $\dfrac{21}{25}$, 0.88, 83%, $\dfrac{17}{20}$

c) 49%, 0.42, $\dfrac{9}{20}$, $\dfrac{23}{50}$, 41%

d) 0.052, 0.6%, $\dfrac{1}{200}$, $\dfrac{29}{500}$, 5%

▶ Fractions of an amount

To find a fraction of an amount:

- divide by the denominator and then multiply by the numerator.

Exercise 3E

1 Calculate:

a) $\frac{1}{3}$ of 18

b) $\frac{1}{5}$ of 70

c) $\frac{1}{4}$ of 96

d) $\frac{1}{7}$ of 245

e) $\frac{1}{6}$ of £342

f) $\frac{1}{8}$ of 904 kg

g) $\frac{1}{20}$ of 600 m

h) $\frac{1}{30}$ of 360 cm

2 Find:

a) $\frac{2}{3}$ of 15

b) $\frac{3}{7}$ of 21

c) $\frac{4}{5}$ of 60

d) $\frac{7}{8}$ of 88

e) $\frac{5}{9}$ of 207

f) $\frac{6}{7}$ of 189

g) $\frac{3}{8}$ of 2000

h) $\frac{1}{4}$ of 10

i) $\frac{2}{5}$ of 16

j) $\frac{3}{8}$ of 20

k) $\frac{3}{20}$ of 36

l) $\frac{7}{30}$ of 24

3 Find the difference between

a) $\frac{3}{4}$ of £16 and $\frac{3}{5}$ of £15

b) $\frac{7}{9}$ of 45 kg and $\frac{2}{3}$ of 42 kg

c) $\frac{5}{8}$ of 60 ml and $\frac{4}{7}$ of 63 ml

d) $\frac{2}{5}$ of 34 miles and $\frac{1}{3}$ of 33 miles

4 a) Increase 63 miles by $\frac{2}{9}$

b) Increase £48 by $\frac{7}{12}$

c) Decrease £65 · 94 by $\frac{3}{7}$

d) Decrease 121 m by $\frac{4}{11}$

5   James takes £2 to his local shop. He spends $\frac{3}{5}$ of his money. How much money does he have left?

6   A film lasts 108 minutes. After $\frac{5}{6}$ of the film, the streaming app fails.

How many minutes of the film have passed before the app failed?

7   In a survey of 154 pupils, $\frac{4}{7}$ of them enjoy doing homework every night.

How many pupils enjoy completing their nightly homework?

8   A standard tin of soup weighs 400 g. A promotional tin advertises $\frac{1}{5}$ extra free. What weight is the promotional tin?

9   A shirt costs £42. It is placed in a sale where all items have $\frac{1}{4}$ off. How much is the shirt in the sale?

10 An estate agent thinks house prices will rise by $\frac{3}{20}$ next month. How much will a house which is

currently valued at £150000 be worth next month if the estate agent is correct?

11 A concert venue can seat 12000 people. One week before a concert, $\frac{7}{8}$ of the tickets had been sold.

a) How many tickets were still available one week before the concert?

Of the seats left, $\frac{4}{5}$ of them were sold with one day to go.

b) If everyone with a ticket attended the concert, how many empty seats were there?

12 A landscape gardener has designed a garden for a client. Their garden is 84 square metres and they requested:

● $\frac{1}{8}$ paving   ● $\frac{5}{8}$ grass   ● $\frac{1}{6}$ bark   ● the remainder to be a vegetable garden.

a) A bag of bark covers 2 square metres of ground. How many bags were required?

b) Turf for the grassed area costs £3 per square metre and can only be bought in whole multiples of a square metre. How much did it cost to turf the grassed area?

c) $\frac{3}{4}$ of the vegetable garden contains potatoes. What area of the total garden is used to grow potatoes?

13 Jez raises money for charity. He gives £150 to his local hospital. This is $\frac{1}{3}$ of the total amount he raised. How much money did he raise altogether?

14 Giuliana has just got her fifth axolotl and now her axolotls make up $\frac{1}{4}$ of her total pets. She is due

to collect a new dog next week. How many pets will Giuliana have altogether after she collects her new dog?

15 240 fans travel to a football match to support their local team. They form $\frac{2}{3}$ of the total number of spectators. How many spectators are there?

16 Ruth spends $\frac{1}{8}$ of her money on a taxi, $\frac{1}{4}$ on food, and $\frac{2}{5}$ when going bowling with friends. She saves the rest of her money. If she spent £20 bowling, how much money did she

a) spend on a taxi

b) spend on food

c) save?

▶ Whole and mixed numbers to improper fractions

A **mixed number** consists of a whole number and a **proper fraction**, for example, $3\frac{1}{2}$ or $5\frac{8}{9}$.

In an **improper fraction** the numerator is larger than or equal to the denominator, for example, $\frac{5}{5}, \frac{4}{3}$ or $\frac{9}{2}$.

● To convert a whole number to an improper fraction, multiply the whole number by the denominator of the fraction.

● For a mixed number, convert the whole number and then add the additional fraction part.

Worked examples

1 Write 1 in thirds

one whole = three thirds

$$= \frac{3}{3}$$

2 How many fifths are in 4?

one whole = five fifths $= \frac{5}{5}$

four whole ones $= \frac{4 \times 5}{5} = \frac{20}{5}$

3 Write these mixed numbers as improper fractions.

a) $3\frac{1}{2}$

$3\frac{1}{2} = \frac{3 \times 2 + 1}{2}$

$= \frac{7}{2}$

b) $4\frac{3}{5}$

$4\frac{3}{5} = \frac{4 \times 5 + 3}{5}$

$= \frac{23}{5}$

c) $2\frac{5}{6}$

$2\frac{5}{6} = \frac{2 \times 6 + 5}{6}$

$= \frac{17}{6}$

d) $3\frac{8}{9}$

$3\frac{8}{9} = \frac{3 \times 9 + 8}{9}$

$= \frac{35}{9}$

Exercise 3F

1 How many quarters are there in each whole number?

   a) 1   b) 6   c) 10   d) 15   e) 22   f) 100   g) 200

2 How many sixths are in each whole number?

   a) 1   b) 4   c) 19   d) 50   e) 72   f) 102   g) 500

3 Convert these mixed numbers into improper fractions.

   a) $4\frac{1}{2}$ b) $2\frac{1}{6}$ c) $3\frac{4}{5}$ d) $1\frac{2}{7}$ e) $8\frac{3}{4}$ f) $6\frac{5}{11}$ g) $7\frac{5}{8}$

   h) $9\frac{7}{12}$ i) $10\frac{2}{5}$ j) $5\frac{9}{10}$ k) $16\frac{1}{4}$ l) $20\frac{2}{7}$ m) $35\frac{1}{3}$ n) $100\frac{12}{13}$

4 Evan is learning to write mixed numbers as improper fractions.

   Identify which of his attempts are correct and which are wrong.

   a) $3\frac{2}{9} = \frac{29}{9}$ b) $4\frac{6}{7} = \frac{34}{7}$ c) $5\frac{2}{9} = \frac{38}{9}$ d) $6\frac{3}{14} = \frac{87}{14}$ e) $7\frac{6}{20} = \frac{73}{10}$

   f) What percentage of questions did Evan get correct?

▶ Improper fractions to whole and mixed numbers

- To convert from an improper fraction to either a whole number or a mixed number, divide the numerator by the denominator to find the whole number.
- Any remainder forms the numerator of the fraction part.

Worked examples Convert to whole or mixed numbers:

a) $\dfrac{14}{2}$ $14 \div 2 = 7$ remainder 0 so $\dfrac{14}{2} = 7$ b) $\dfrac{64}{10}$ $64 \div 10 = 6$ remainder 4 so $\dfrac{64}{10} = 6\dfrac{4}{10} = 6\dfrac{2}{5}$

Exercise 3G

1   Convert these improper fractions to whole numbers.

a) $\dfrac{12}{2}$ b) $\dfrac{36}{9}$ c) $\dfrac{48}{4}$ d) $\dfrac{21}{3}$ e) $\dfrac{160}{10}$ f) $\dfrac{480}{20}$ g) $\dfrac{5600}{70}$

2   Write these improper fractions as mixed numbers.

a) $\dfrac{17}{2}$ b) $\dfrac{11}{3}$ c) $\dfrac{24}{5}$ d) $\dfrac{19}{4}$ e) $\dfrac{33}{10}$ f) $\dfrac{49}{8}$ g) $\dfrac{77}{6}$

h) $\dfrac{22}{4}$ i) $\dfrac{26}{6}$ j) $\dfrac{57}{9}$ k) $\dfrac{46}{8}$ l) $\dfrac{76}{10}$ m) $\dfrac{122}{4}$ n) $\dfrac{98}{20}$

3   Convert these improper fractions to mixed numbers and write your answers from smallest to largest.

$\dfrac{44}{7}$ $\dfrac{25}{4}$ $\dfrac{20}{3}$ $\dfrac{34}{5}$ $\dfrac{46}{7}$ $\dfrac{13}{2}$ $\dfrac{31}{5}$

4   a) Place the cards in the correct place so that the numbers beside each other from one card to the next are always matching. One card does not fit. Which is it?

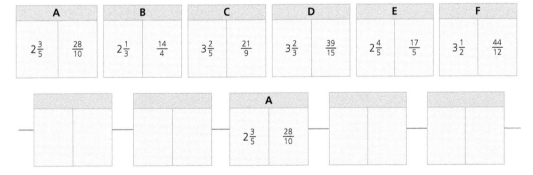

b) Find another combination starting with card F in the middle.

c) Is there any card that, if put in the middle, the other cards can't be placed? Can you spot a pattern?

▶ Addition and subtraction of fractions: common denominators

The denominators of both fractions must be the same before you can add or subtract.

$$\frac{1}{3} + \frac{1}{3} = \frac{2}{3}$$

One-third add one-third gives two-thirds.

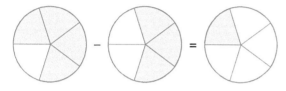

$$\frac{4}{5} - \frac{3}{5} = \frac{1}{5}$$

Four-fifths subtract three-fifths equals one-fifth.

- Add or subtract the numerators.
- The denominator remains unchanged.

Worked examples

1
$$\frac{5}{9} + \frac{2}{9}$$
$$= \frac{7}{9}$$

2
$$\frac{8}{13} - \frac{3}{13}$$
$$= \frac{5}{13}$$

3
$$\frac{1}{8} + \frac{3}{8}$$
$$= \frac{4}{8}$$
$$= \frac{1}{2}$$

4
$$\frac{9}{10} - \frac{5}{10}$$
$$= \frac{4}{10}$$
$$= \frac{2}{5}$$

5
$$\frac{11}{16} + \frac{5}{16}$$
$$= \frac{16}{16}$$
$$= 1$$

6
$$\frac{6}{7} + \frac{8}{7} - \frac{2}{7}$$
$$= \frac{12}{7}$$
$$= 1\frac{5}{7}$$

Exercise 3H

1 Simplify:

a) $\frac{1}{7} + \frac{3}{7}$

b) $\frac{2}{5} + \frac{1}{5}$

c) $\frac{4}{9} + \frac{2}{9}$

d) $\frac{1}{8} + \frac{5}{8}$

e) $\frac{4}{15} + \frac{8}{15}$

f) $\frac{3}{8} + \frac{5}{8}$

g) $\frac{8}{9} - \frac{3}{9}$

h) $\frac{3}{5} - \frac{2}{5}$

i) $\frac{7}{8} - \frac{3}{8}$

j) $\frac{8}{9} - \frac{2}{9}$

k) $\frac{11}{12} - \frac{7}{12}$

l) $\frac{17}{20} - \frac{9}{20}$

2 Calculate, giving your answer as a mixed number:

a) $\frac{4}{5} + \frac{3}{5}$

b) $\frac{7}{11} + \frac{9}{11}$

c) $\frac{23}{15} - \frac{4}{15}$

d) $\frac{37}{9} - \frac{14}{9}$

e) $\frac{7}{8} + \frac{15}{8}$

f) $\frac{43}{10} - \frac{19}{10}$

→

3 Simplify, writing your answer as a mixed number where possible:

a) $\dfrac{1}{13} + \dfrac{4}{13} + \dfrac{6}{13}$ b) $\dfrac{4}{11} - \dfrac{2}{11} + \dfrac{18}{11}$ c) $\dfrac{5}{9} + \dfrac{14}{9} - \dfrac{1}{9}$ d) $\dfrac{12}{15} - \dfrac{1}{15} + \dfrac{16}{15} - \dfrac{7}{15}$

4 On an allotment, $\dfrac{5}{9}$ of the land is used for potatoes, $\dfrac{1}{9}$ for turnips and the remainder for carrots.

What fraction of the allotment is used for: a) potatoes and turnips?   b) carrots?

5 During a sports day, $\dfrac{1}{7}$ of pupils chose to play hockey, $\dfrac{4}{7}$ to play football and $\dfrac{2}{7}$ chose athletics.

a) What fraction of pupils chose either football or athletics?

b) What fraction more chose to play football than hockey?

c) If there were 280 pupils in the year, how many more pupils chose to play football than hockey?

6 There are four different colours of car in a showroom. $\dfrac{9}{20}$ are black, $\dfrac{7}{20}$ are blue, $\dfrac{3}{20}$ are white and $\dfrac{1}{20}$ are red.

a) What fraction of the cars in the showroom are either black or red?

b) As a fraction, how many more cars in the showroom are blue than white?

c) As a fraction, how many more cars in the showroom are black or blue than red?

7 Find two fractions which:

a) add to give $\dfrac{9}{11}$ and have a difference of $\dfrac{1}{11}$

b) add to give 1 and have a difference of $\dfrac{3}{5}$

8 In a magic square, all the rows, columns and diagonals add to give the same number.

Copy and complete these magic squares.

a)

		$\dfrac{2}{13}$
	$\dfrac{4}{13}$	
$\dfrac{6}{13}$		$\dfrac{5}{13}$

b)

	$\dfrac{5}{19}$	
$\dfrac{4}{19}$	$\dfrac{3}{19}$	$\dfrac{8}{19}$

c)

$\dfrac{8}{21}$	$\dfrac{1}{7}$	
	$\dfrac{5}{21}$	
		$\dfrac{2}{21}$

d)

		$\dfrac{1}{9}$
	$\dfrac{4}{9}$	$\dfrac{2}{3}$
		$\dfrac{5}{9}$

▶ Addition and subtraction of fractions: different denominators

The denominators of the fractions must be the same before adding or subtracting.

- Identify the lowest common multiple (LCM) of the denominators.
- Find equivalent fractions with this lowest common multiple as each denominator.
- Add or subtract and then simplify if possible.

$$\frac{1^{\times 4}}{2_{\times 4}} + \frac{3}{8}$$

$$= \frac{4}{8} + \frac{3}{8}$$

$$= \frac{7}{8}$$

The LCM of 2 and 8 is 8 so multiply the numerator and denominator of the first fraction by 4 to get a common denominator of 8 in both fractions.

$$\frac{2^{\times 4}}{3_{\times 4}} - \frac{1^{\times 3}}{4_{\times 3}}$$

$$= \frac{8}{12} - \frac{3}{12}$$

$$= \frac{5}{12}$$

The LCM of 3 and 4 is 12 so multiply the numerator and denominator of the first fraction by 4 and the numerator and denominator of the second fraction by 3.

Worked examples

1 $\frac{1}{9} + \frac{2}{3}$ (LCM = 9)

$$= \frac{1}{9} + \frac{2^{\times 3}}{3_{\times 3}}$$

$$= \frac{1}{9} + \frac{6}{9}$$

$$= \frac{7}{9}$$

2 $\frac{4}{5} - \frac{1}{2}$ (LCM = 10)

$$= \frac{4^{\times 2}}{5_{\times 2}} - \frac{1^{\times 5}}{2_{\times 5}}$$

$$= \frac{8}{10} - \frac{5}{10}$$

$$= \frac{3}{10}$$

3 $\frac{3}{4} + \frac{5}{6}$ (LCM = 12)

$$= \frac{3^{\times 3}}{4_{\times 3}} + \frac{5^{\times 2}}{6_{\times 2}}$$

$$= \frac{9}{12} + \frac{10}{12}$$

$$= \frac{19}{12}$$

$$= 1\frac{7}{12}$$

4 $\frac{1}{3} + \frac{3}{5} - \frac{1}{10}$ (LCM = 30)

$$= \frac{1^{\times 10}}{3_{\times 10}} + \frac{3^{\times 6}}{5_{\times 6}} - \frac{1^{\times 3}}{10_{\times 3}}$$

$$= \frac{10}{30} + \frac{18}{30} - \frac{3}{30}$$

$$= \frac{25}{30}$$

$$= \frac{5}{6}$$

Exercise 3I

1 Find the lowest common multiple of the numbers in each group.

a) 4 and 12   b) 6 and 7   c) 2 and 9   d) 2, 4 and 6   e) 3, 7 and 10

2 Find the lowest common multiple of the denominators of the fractions in each group.

a) $\frac{3}{5}$ and $\frac{7}{10}$ b) $\frac{1}{2}$ and $\frac{2}{3}$ c) $\frac{5}{6}$ and $\frac{3}{10}$ d) $\frac{2}{9}$ and $\frac{1}{6}$ e) $\frac{5}{7}, \frac{2}{3}$ and $\frac{1}{2}$ ➡

3 Copy and complete to make equivalent fractions.

a) $\dfrac{1}{2} = \dfrac{}{6}$ b) $\dfrac{3}{4} = \dfrac{15}{}$ c) $\dfrac{6}{7} = \dfrac{}{63}$ d) $\dfrac{4}{5} = \dfrac{48}{}$ e) $\dfrac{6}{11} = \dfrac{24}{}$ f) $\dfrac{16}{21} = \dfrac{}{105}$

4 Find, writing your answer as a mixed number where possible:

a) $\dfrac{1}{2} + \dfrac{1}{4}$ b) $\dfrac{2}{3} + \dfrac{1}{12}$ c) $\dfrac{4}{15} + \dfrac{2}{5}$ d) $\dfrac{9}{14} + \dfrac{6}{7}$ e) $\dfrac{2}{3} + \dfrac{1}{2}$ f) $\dfrac{4}{9} + \dfrac{1}{4}$

g) $\dfrac{3}{7} + \dfrac{5}{6}$ h) $\dfrac{7}{12} + \dfrac{4}{5}$ i) $\dfrac{9}{10} + \dfrac{7}{8}$ j) $\dfrac{2}{5} + \dfrac{3}{16}$ k) $\dfrac{5}{6} + \dfrac{2}{13}$ l) $\dfrac{17}{20} + \dfrac{29}{30}$

5 Simplify:

a) $\dfrac{7}{8} - \dfrac{1}{4}$ b) $\dfrac{9}{10} - \dfrac{3}{5}$ c) $\dfrac{2}{3} - \dfrac{1}{9}$ d) $\dfrac{5}{8} - \dfrac{1}{2}$ e) $\dfrac{6}{7} - \dfrac{2}{3}$ f) $\dfrac{19}{20} - \dfrac{5}{7}$

6 Find, writing your answer as a mixed number where possible:

a) $\dfrac{5}{6} + \dfrac{1}{3} + \dfrac{1}{2}$ b) $\dfrac{1}{2} + \dfrac{2}{3} - \dfrac{4}{5}$ c) $\dfrac{1}{4} + \dfrac{3}{8} - \dfrac{2}{5}$ d) $\dfrac{7}{9} - \dfrac{2}{3} + \dfrac{1}{2}$

e) $\dfrac{3}{5} - \dfrac{1}{4} - \dfrac{1}{8}$ f) $\dfrac{2}{7} + \dfrac{2}{3} - \dfrac{9}{10}$ g) $\dfrac{7}{20} - \dfrac{1}{30} + \dfrac{16}{40}$ h) $\dfrac{5}{6} + \dfrac{2}{3} - \dfrac{3}{2}$

7 In a supermarket aisle, $\dfrac{2}{5}$ of the produce is yoghurt, $\dfrac{1}{4}$ is cheese, $\dfrac{1}{6}$ is butter and the remainder is milk.

What fraction of the aisle contains:

a) yoghurt and cheese

b) cheese and butter

c) milk?

8 A young enterprise group are selling merchandise for charity.

Their initial stock is made up of $\dfrac{1}{4}$ stickers, $\dfrac{2}{7}$ badges, $\dfrac{4}{9}$ pens and the rest is lanyards.

a) What fraction of their stock do the pens and badges make?

b) What fraction more of their stock is pens compared to stickers?

c) If the group sell all their badges and lanyards but nothing else, what fraction of the initial stock is left?

9 Find two fractions that:

a) add to give $\dfrac{3}{4}$ and have a difference of $\dfrac{1}{12}$

b) add to give $\dfrac{17}{15}$ and have a difference of $\dfrac{1}{5}$

c) add to give $1\dfrac{3}{10}$ and have a difference of $\dfrac{1}{2}$

10 There are two solutions for both of these questions. Find three positive fractions that:

a) add to give $\dfrac{5}{3}$, one fraction is $\dfrac{4}{9}$ and two fractions are equal

b) add to give $1\dfrac{4}{15}$, one fraction is $\dfrac{2}{3}$ and one fraction is the double of another.

▶ Multiplying and dividing fractions

You can write the **product** of two fractions as a single fraction by multiplying the numerators together and multiplying the denominators together. For example, $\frac{2}{5} \times \frac{11}{8} = \frac{2 \times 11}{5 \times 8}$.

Therefore, when multiplying fractions, make your calculation easier by simplifying any number on a numerator with any number on a denominator. Note that the word 'of' refers to multiplication in fraction calculations.

- If possible, simplify numbers on the numerators with numbers on the denominators.
- Multiply numerators together and denominators together.

Worked examples

1
$$\frac{1}{3} \times \frac{2}{5}$$
$$= \frac{2}{15}$$
There are no numbers that simplify in this example.

2
$$\frac{4}{5} \times \frac{10}{11}$$
$$= \frac{4}{\cancel{5}_{1}} \times \frac{\cancel{10}^{2}}{11}$$
$$= \frac{8}{11}$$

3
$$\frac{2}{3} \text{ of } \frac{9}{14}$$
$$= \frac{2}{3} \times \frac{9}{14}$$
$$= \frac{\cancel{2}^{1}}{\cancel{3}_{1}} \times \frac{\cancel{9}^{3}}{\cancel{14}_{7}}$$
$$= \frac{3}{7}$$

4
$$6 \times \frac{7}{9}$$
$$= \frac{6}{1} \times \frac{7}{9}$$
$$= \frac{\cancel{6}^{2}}{1} \times \frac{7}{\cancel{9}_{3}}$$
$$= \frac{14}{3}$$
$$= 4\frac{2}{3}$$

Any whole number can be written as a fraction by dividing it by 1.

Exercise 3J

Throughout this exercise, write any answer that is an improper fraction as a mixed number.

1 Multiply (there is no simplification possible before multiplying):

a) $\frac{1}{2} \times \frac{1}{3}$
b) $\frac{1}{6} \times \frac{7}{8}$
c) $\frac{3}{7} \times \frac{2}{11}$
d) $\frac{5}{2} \times \frac{9}{13}$
e) $\frac{5}{6} \times \frac{13}{9}$
f) $\frac{6}{5} \times \frac{14}{13}$

2 Find (remember to simplify first):

a) $\frac{1}{2} \times \frac{4}{5}$
b) $\frac{1}{6} \times \frac{2}{7}$
c) $\frac{10}{3} \times \frac{1}{5}$
d) $\frac{1}{4} \times \frac{6}{5}$
e) $\frac{4}{9} \times \frac{6}{7}$
f) $\frac{7}{8} \times \frac{20}{3}$

3 Calculate:

a) $\frac{2}{3} \times \frac{9}{10}$
b) $\frac{2}{9} \times \frac{3}{4}$
c) $\frac{5}{6} \times \frac{3}{10}$
d) $\frac{7}{8} \times \frac{2}{21}$
e) $\frac{3}{5} \times \frac{35}{18}$
f) $\frac{8}{11} \times \frac{55}{6}$

4 Find:

a) $4 \times \frac{2}{3}$
b) $6 \times \frac{3}{5}$
c) $8 \times \frac{3}{4}$
d) $\frac{5}{6} \times 3$
e) $\frac{19}{15} \times 10$
f) $\frac{3}{8} \times 12$

5 In a class, $\frac{4}{5}$ of the pupils play a musical instrument. From these pupils, $\frac{2}{3}$ of them play the piano.

What fraction of the class play the piano?

6 In a tombola, $\frac{3}{4}$ of the donations are edible. Of the edible donations, $\frac{8}{9}$ contain chocolate.

What fraction of the total donations do not contain chocolate?

7 Fill in the blanks to make the following statements true.

a) $\frac{1}{3} \times \frac{2}{4} \times \frac{3}{2} \times \boxed{} = 1$

b) $\frac{4}{5} \times \frac{1}{2} \times \frac{1}{4} \times \boxed{} = \frac{3}{20}$

c) $7 \times \frac{2}{5} \times \frac{1}{4} = \frac{1}{5} \times \boxed{}$

d) $\frac{2}{3} \times \frac{4}{5} \times \frac{5}{6} = \frac{1}{4} \times 2 \times \boxed{}$

Consider $10 \div 2$. You can write this as the multiplication $\frac{1}{2}$ of 10 or $\frac{1}{2} \times 10$.

Similarly, $12 \div 3$ is the same as $\frac{1}{3}$ of 12 or $12 \times \frac{1}{3}$.

The multiplying fraction $\frac{1}{2}$ is the reciprocal of the divisor 2 and $\frac{1}{3}$ is the reciprocal of 3.

Now consider $4 \div \frac{2}{3}$. How many $\frac{2}{3}$ are there in 4?

$$4 = \frac{12}{3} = \frac{2}{3} + \frac{2}{3} + \frac{2}{3} + \frac{2}{3} + \frac{2}{3} + \frac{2}{3}$$

There are 6 lots of $\frac{2}{3}$ in 4 therefore $4 \div \frac{2}{3} = 6$. We can get this by rewriting the original division as a multiplication with the reciprocal of the dividing fraction.

$$4 \div \frac{2}{3} = 4 \times \frac{3}{2}$$
$$= \frac{12}{2}$$
$$= 6$$

To find the reciprocal of a fraction, interchange the numerator and denominator. Here, the reciprocal of $\frac{2}{3}$ is $\frac{3}{2}$.

So to divide by a fraction we change the calculation to a multiplication?

Yes but remember to multiply by the reciprocal of the dividing fraction.

To get the reciprocal of the fraction, swap the numerator and denominator.

To divide by a fraction:

- Keep the first number unchanged and **multiply** by the reciprocal of the dividing fraction.
- Complete by using the technique of multiplying fractions.
- Do not try to simplify until you have written the calculation as a multiplication.

Worked examples

1. $\dfrac{1}{4} \div \dfrac{2}{5}$

$= \dfrac{1}{4} \times \dfrac{5}{2}$

$= \dfrac{5}{8}$

2. $\dfrac{2}{3} \div \dfrac{4}{7}$

$= \dfrac{\overset{1}{2}}{3} \times \dfrac{7}{\underset{2}{4}}$

$= \dfrac{7}{6}$

$= 1\dfrac{1}{6}$

3. $\dfrac{5}{6} \div \dfrac{35}{24}$

$= \dfrac{\overset{1}{5}}{\underset{1}{6}} \times \dfrac{\overset{4}{24}}{\underset{7}{35}}$

$= \dfrac{4}{7}$

4. $6 \div \dfrac{3}{8}$

$= \dfrac{\overset{2}{6}}{1} \times \dfrac{8}{\underset{1}{3}}$

$= \dfrac{16}{1}$

$= 16$

5. $\dfrac{2}{9} \div 16$

$= \dfrac{\overset{1}{2}}{9} \times \dfrac{1}{\underset{8}{16}}$

$= \dfrac{1}{72}$

Exercise 3K

Throughout this exercise, write any answer that is an improper fraction as a mixed number.

1. Find:

   a) $\dfrac{1}{4} \div \dfrac{1}{3}$
   b) $\dfrac{3}{7} \div \dfrac{1}{2}$
   c) $\dfrac{1}{6} \div \dfrac{3}{5}$
   d) $\dfrac{5}{9} \div \dfrac{4}{5}$
   e) $\dfrac{2}{7} \div \dfrac{1}{9}$
   f) $\dfrac{3}{2} \div \dfrac{5}{11}$

2. Calculate:

   a) $\dfrac{1}{2} \div \dfrac{7}{8}$
   b) $\dfrac{1}{6} \div \dfrac{3}{4}$
   c) $\dfrac{5}{7} \div \dfrac{10}{9}$
   d) $\dfrac{2}{5} \div \dfrac{6}{7}$
   e) $\dfrac{3}{8} \div \dfrac{1}{4}$
   f) $\dfrac{16}{5} \div \dfrac{7}{5}$

3. Simplify:

   a) $\dfrac{3}{4} \div \dfrac{9}{8}$
   b) $\dfrac{2}{3} \div \dfrac{16}{21}$
   c) $\dfrac{3}{8} \div \dfrac{15}{16}$
   d) $\dfrac{7}{2} \div \dfrac{21}{10}$
   e) $\dfrac{4}{5} \div \dfrac{8}{15}$
   f) $\dfrac{3}{8} \div \dfrac{9}{40}$

4. Simplify:

   a) $4 \div \dfrac{2}{3}$
   b) $2 \div \dfrac{3}{5}$
   c) $12 \div \dfrac{8}{9}$
   d) $\dfrac{2}{7} \div 6$
   e) $\dfrac{5}{8} \div 10$
   f) $\dfrac{9}{2} \div 15$

5. The area of a rectangle is $14\,\text{cm}^2$. If the length is $\dfrac{8}{5}\,\text{cm}$, what is the breadth of the rectangle?

6. How many lengths of $\dfrac{5}{4}\,\text{m}$ of tape can be cut from a one-kilometre roll?

7. How many full $\dfrac{2}{3}\,\text{m}$ lengths of wire can be cut from $45\,\text{m}$?

8. Find:

   a) $\dfrac{2}{3}$ of $\left(20 \div \dfrac{1}{6}\right)$
   b) $\left(\dfrac{1}{3} + \dfrac{6}{7}\right) \div \dfrac{5}{3}$
   c) $2 + \dfrac{4}{9} \div 8$
   d) $1 + \dfrac{7}{8} \div \dfrac{5}{2} \times 30$

Percentages of an amount: with a calculator

- Using a calculator we can calculate the percentage of an amount as $\dfrac{\text{percentage}}{100} \times \text{amount}$.

- Do not round any decimal values until your final answer. Write your unrounded answer before rounding.

Worked examples

1 Calculate 27% of 312 ml

$\dfrac{27}{100} \times 312 = 27 \div 100 \times 312$

$= 84 \cdot 24 \text{ ml}$

2 Decrease 12 kg by 6·5%

$\dfrac{6 \cdot 5}{100} \times 12 = 6 \cdot 5 \div 100 \times 12$

$= 0 \cdot 78 \text{ kg}$

$12 - 0 \cdot 78 = 11 \cdot 22 \text{ kg}$

3 Find 0·32% of £619·50

$\dfrac{0 \cdot 32}{100} \times 619 \cdot 5 = 0 \cdot 32 \div 100 \times 619 \cdot 5$

$= 1 \cdot 9824$

$= £1 \cdot 98$

Remember: always round any money values to 2 decimal places.

Exercise 3L

1 Calculate:

  a) 43% of 25   **b)** 18% of 74   **c)** 72% of 652   **d)** 56% of 830

2 Find:

  a) 3·2% of 400   **b)** 47·1% of 250   **c)** 66·4% of 190   **d)** 27·4% of 376

  e) 93·5% of 682   **f)** 1·14% of 60   **g)** 2·76% of 82   **h)** 0·56% of 30

3 How much will a £4·50 train fare be after a 3·5% price increase?

4 A bike shop has a 12·5% sale on all goods. If a bike normally costs £99, how much will it cost in the sale?

5 After a strict training regime, Jon beats his previous personal best running time of 68·2 seconds by 1·3%.

  a) How much faster was Jon? **b)** To one decimal place, what is his new personal best?

6 Alba Bank offers a savings account with an interest rate of 2·9% per annum. A similar savings account with Caledonia Bank offers 2·71% per annum as an interest rate.

  Luna has £2500 to put into an account. Assuming she takes no money out, how much

  a) would Luna have in Alba Bank after one year?

  b) would Luna have in Caledonia Bank after one year?

  c) more interest would she make with Alba Bank?

7 A car decreases in value by:
- 8% of its original value in the first year
- 6·5% of its value at the end of the first year in the second year
- 4·2% of its value at the end of the second year in the third year.

  If the car originally cost £12500,

  a) how much is it worth after 3 years **b)** what value has the car lost after 3 years?

▶ Percentages of an amount: without a calculator

Recall the following links between percentages and fractions alongside the associated calculations.

Percentage	Fraction	Calculation
1%	$\frac{1}{100}$	divide by 100
10%	$\frac{10}{100} = \frac{1}{10}$	divide by 10
25%	$\frac{25}{100} = \frac{1}{4}$	divide by 4
50%	$\frac{50}{100} = \frac{1}{2}$	divide by 2
75%	$\frac{75}{100} = \frac{3}{4}$	divide by 4 and then multiply by 3

● To find a percentage of an amount without a calculator, use multiples of the percentages in this table.

For example, to find 32%:

→ Find 10% and multiply it by 3 to get 30%

→ Find 1% and multiply it by 2 to get 2%

→ Add your answers for 30% and 2% together to get 32%.

Worked examples Find:

1 7% of £200

Find 1%: $1\% = 200 \div 100$
    $= 2$

Now multiply by 7 to get 7%:
    $7\% = 7 \times 1\%$
    $= 7 \times 2$
    $= £14$

2 51% of 3 kg

Find 50%: $50\% = 3 \div 2$
    $= 1{\cdot}5$

Find 1%: $1\% = 3 \div 100$
    $= 0{\cdot}03$

Adding gives $51\% = 50\% + 1\%$
    $= 1{\cdot}5 + 0{\cdot}03$
    $= 1{\cdot}53$ kg

3 43% of 620 ml

$10\% = 620 \div 10$ $40\% = 4 \times 10\%$
    $= 62$ ⟶ $= 4 \times 62$
    $= 248$

$1\% = 620 \div 100$ $3\% = 3 \times 1\%$
    $= 6{\cdot}2$ ⟶ $= 3 \times 6{\cdot}2$
    $= 18{\cdot}6$

Adding gives $43\% = 40\% + 3\%$
    $= 248 + 18{\cdot}6$
    $= 266{\cdot}6$ ml

4 82% of 90 g

$10\% = 90 \div 10$ $80\% = 8 \times 10\%$
    $= 9$ ⟶ $= 8 \times 9$
    $= 72$

$1\% = 90 \div 100$ $2\% = 2 \times 1\%$
    $= 0{\cdot}9$ ⟶ $= 2 \times 0{\cdot}9$
    $= 1{\cdot}8$

Adding gives $82\% = 80\% + 2\%$
    $= 72 + 1{\cdot}8$
    $= 73{\cdot}8$ g

Exercise 3M

1   Find 1% first and use it to calculate:

   a) 2% of 400   b) 3% of 900   c) 4% of 500   d) 5% of 800

   e) 6% of 30   f) 7% of 60   g) 8% of 70   h) 9% of 5

2   Find 10% first and use it to calculate:

   a) 40% of 200   b) 70% of 900   c) 60% of 300   d) 20% of 800

   e) 90% of 50   f) 30% of 40   g) 80% of 60   h) 40% of 9

3   In a sale, the price of a console game is advertised as 20% off. It originally cost £40.
    What does the game cost in the sale?

4   A standard pack of grapes weighs 200 g. As part of a special offer, the pack has 30% extra free.
    How much does the special offer pack weigh?

5   Copy and complete:

   a) $23\% = 2 \times 10\% + \boxed{} \times 1\%$ b) $87\% = \boxed{} \times 10\% + 7 \times 1\%$

   c) $69\% = \boxed{} \times 10\% + \boxed{} \times 1\%$ d) $54\% = 5 \times \boxed{} + 4 \times \boxed{}$

6   Calculate:

   a) 24% of 500   b) 37% of 800   c) 52% of 600   d) 65% of 400

   e) 81% of 700   f) 49% of 300   g) 96% of 200   h) 92% of 500

7   37% of people who attended a concert bought t-shirts.
    If there were 12 000 people at the concert, how many bought t-shirts?

8   Calculate:

   a) 41% of 60   b) 26% of 90   c) 73% of 40   d) 58% of 20

   e) 39% of 70   f) 84% of 10   g) 6% of 4   h) 28% of 8

   i) 71% of 6   j) 63% of 5   k) 94% of 3   l) 88% of 7

9   In a local election, 620 people voted. The winning candidate gained 45% of the votes.
    How many votes did they get?

10 In a football game, 1450 successful passes were made. The winning team made 68% of the passes.
    How many passes did they make?

11 It is recommended that men should consume no more than 30 g of saturated fat a day. A takeaway
    meal weighs 720 g and saturated fat makes up 4·5% of the meal.
    Would this takeaway be within healthy limits for the day? Justify your answer.

Check-up 👍

1 Simplify, writing your answer as a mixed number where possible:

   a) $\dfrac{4}{10}$ b) $\dfrac{12}{16}$ c) $\dfrac{10}{30}$ d) $\dfrac{21}{28}$ e) $\dfrac{45}{20}$ f) $\dfrac{99}{18}$ g) $\dfrac{62}{40}$

2 From 100 people at the cinema, 26 bought popcorn. What fraction of the people bought popcorn?

3 Convert to a fraction:

   a) 0·8   b) 0·26   c) 0·86   d) 0·77   e) 0·105   f) 0·052   g) 0·006

4 Write each decimal in question 3 as a percentage.

5 Convert to a fraction:

   a) 17%   b) 45%   c) 51%   d) 6%   e) 94%   f) 72%   g) 62%

6 Write each percentage in question 5 as a decimal.

7 Convert to a decimal:

   a) $\dfrac{7}{10}$ b) $\dfrac{2}{5}$ c) $\dfrac{24}{40}$ d) $\dfrac{5}{8}$ e) $\dfrac{9}{20}$ f) $\dfrac{7}{50}$ g) $\dfrac{41}{80}$

8 Write each fraction in question 7 as a percentage.

9 Write in order, from smallest to largest:

   a) 0·32, 34%, $\dfrac{31}{100}$, 0·3, $\dfrac{7}{20}$ b) $\dfrac{3}{4}$, 77%, $\dfrac{37}{50}$, 0·73, $\dfrac{19}{25}$

10 Calculate:

   a) $\dfrac{1}{3}$ of 360°   b) $\dfrac{3}{5}$ of 60 m   c) $\dfrac{5}{6}$ of 312 ml

   d) $\dfrac{4}{7}$ of 763 kg   e) $\dfrac{2}{9}$ of 450 miles   f) $\dfrac{7}{8}$ of £638

11 From 440 smoothies sold in a cafe, $\dfrac{3}{4}$ had strawberries in them, $\dfrac{1}{5}$ had blueberries and $\dfrac{5}{8}$ had bananas. How many

   a) smoothies had banana in them

   b) more smoothies had strawberries in them than blueberries?

12 Convert to an improper fraction:

   a) $7\dfrac{1}{2}$ b) $5\dfrac{2}{3}$ c) $3\dfrac{7}{9}$ d) $4\dfrac{5}{8}$ e) $6\dfrac{9}{10}$ f) $12\dfrac{3}{4}$

13 Write as a mixed number:

   a) $\dfrac{14}{3}$ b) $\dfrac{18}{5}$ c) $\dfrac{47}{7}$ d) $\dfrac{48}{9}$ e) $\dfrac{52}{8}$ f) $\dfrac{76}{6}$

14 Calculate and simplify where possible:

   a) $\dfrac{1}{4} + \dfrac{1}{4}$ b) $\dfrac{2}{9} + \dfrac{4}{9}$ c) $\dfrac{8}{15} + \dfrac{7}{15}$ d) $\dfrac{7}{8} - \dfrac{1}{8}$ e) $\dfrac{9}{10} - \dfrac{5}{10}$ f) $\dfrac{5}{12} + \dfrac{9}{12} - \dfrac{7}{12}$

15 Simplify, writing your answer as a mixed number where possible.

   a) $\dfrac{1}{4} + \dfrac{1}{3}$ b) $\dfrac{7}{8} + \dfrac{3}{4}$ c) $\dfrac{7}{10} + \dfrac{2}{5}$ d) $\dfrac{3}{4} + \dfrac{2}{3}$ e) $\dfrac{3}{8} + \dfrac{5}{6}$ f) $\dfrac{4}{5} + \dfrac{8}{9}$

   g) $\dfrac{7}{10} - \dfrac{1}{5}$ h) $\dfrac{11}{12} - \dfrac{3}{4}$ i) $\dfrac{13}{18} - \dfrac{1}{3}$ j) $\dfrac{7}{8} - \dfrac{2}{3}$ k) $\dfrac{6}{7} - \dfrac{4}{5}$ l) $\dfrac{11}{15} - \dfrac{7}{10}$ →

16 On Monday, Dougie cut $\frac{1}{5}$ of his grass before the rain stopped him. On Tuesday, he managed to cut another $\frac{1}{3}$ before the rain started.

   a) What fraction of his grass did Dougie manage to cut over Monday and Tuesday?

   b) His son claimed he hadn't managed to cut half of the grass. Was his son correct? Explain your answer.

17 When splitting up chores, Judy agreed to do $\frac{5}{8}$ of them, Andy will do $\frac{1}{3}$ and Scott will complete the rest.

   a) What fraction of the chores will Scott do?

   b) What fraction more of the chores will Judy do than to Andy?

18 At an art exhibition, $\frac{1}{7}$ of the paintings were oil, $\frac{3}{14}$ were watercolour, $\frac{1}{2}$ were acrylic and the remainder were pastel.

   a) What fraction of the paintings were oil or watercolour?

   b) What fraction was completed in pastel?

   c) What fraction more were acrylic compared to pastel paintings?

19 Find:

   a) $\frac{1}{7} \times \frac{1}{5}$ b) $\frac{1}{2} \times \frac{6}{7}$ c) $\frac{3}{4} \times \frac{5}{9}$ d) $\frac{4}{3} \times \frac{9}{16}$ e) $\frac{8}{9} \times \frac{15}{56}$ f) $\frac{3}{10} \times \frac{50}{21}$

20 Simplify, writing your answer as a mixed number where possible:

   a) $\frac{7}{9} \div \frac{2}{5}$ b) $\frac{3}{4} \div \frac{6}{7}$ c) $\frac{11}{12} \div \frac{2}{3}$ d) $\frac{9}{16} \div \frac{3}{2}$ e) $\frac{9}{4} \div \frac{15}{8}$ f) $\frac{28}{9} \div \frac{14}{3}$

21 Richard collects half of his wages early as he is going on holiday. He gives one-eighth of those wages to his friend. What fraction of his total wages did he give to his friend?

22 How many full $\frac{1}{3}$ litre glasses can be filled from a $\frac{9}{2}$ litre carton of milk?

23 Use a calculator to increase £14 by 7·6%.

24 Without a calculator, find:

   a) 30% of 60 g   b) 5% of 200 mm   c) 14% of 350 km

   d) 56% of £405   e) 84% of £338   f) 72% of £31·50

25 A holiday company advertises a sale offering 24% off for 24 hours.

   If the holiday normally costs £1435, how much will it cost in the sale?

26 Prize money is shared between three people.

   First prize is awarded 45% of the money and third prize receives $\frac{3}{20}$ of the total prize fund.

   a) What percentage was awarded to second prize?

   The total prize money is £75.

   b) How much more does first prize get than third prize?

4 Ratio and proportion

▶ Expressing a ratio in its simplest form

We use **ratio** to compare quantities and describe the relationship between them.

- List the number of each item under the correct heading.
- Divide each item by the highest common factor.

Worked examples

1 A logo has 6 stars and 3 arrows. Express the ratio of stars to arrows in its **simplest form**.

stars : arrows

6:3 ÷3

÷3

2:1

The ratio of stars to arrows is 2:1.
There are two stars for each arrow.

2 One Scottish spring had 12 dry, 48 sunny and 36 wet days. Express the ratio of dry to sunny to wet days in its simplest form.

dry : sunny : wet

12: 48 :36

÷12 ( ) ÷12

1: 4 :3

dry : sunny : wet

÷2 (12: 48 :36) ÷2

÷6 (6: 24 :18) ÷6

1: 4 :3

Both solutions are correct.
If you do not spot the highest common factor, take extra steps to divide out remaining factors.

3 At four weeks old a terrier pup weighs 300 g. As an adult it grows to 2·7 kg. Express the ratio in its simplest form.

2·7 kg
= 2·7 × 1000
= 2700 g

pup : adult

300:2700

÷300 ( ) ÷300

1:9

The units must be the same before we can compare.

Remember, to divide by 300, divide by 3 and divide by 100.

Exercise 4A

1 Express each ratio in its simplest form:
- a) 10:15
- b) 16:8
- c) 16:12
- d) 18:45
- e) 10:40
- f) 24:54
- g) 12:24
- h) 72:64

2
- a) 28:70
- b) 99:44
- c) 45:60
- d) 56:64
- e) 360:80
- f) 210:420
- g) 490:140
- h) 300:540

3 Write each ratio under appropriate headings, then express the ratio in its simplest form.
- a) There are 4000 sheep and 400 cattle grazing on Scottish nature reserves.
- b) A herd of grey seals has 170 adults and 85 pups.

➜

c) A flock of grouse has 14 adults and 70 chicks.

d) A town square has 420 pigeons and 192 crows.

Summer Hit	Fruit Treat
Pineapple 160 ml	Apple 180 ml
Mango 120 ml	Orange 45 ml

4 The table shows recipes for drinks.
Express each recipe as a ratio in its simplest form.

5 Express each ratio in simplest form:

a) $16:24:40$ b) $35:42:56$ c) $18:9:90$ d) $33:15:42$

e) $56:21:105$ f) $75:225:600$ g) $510:34:170$ h) $96:129:78$

6 Here are two recipes for making scones. **Recipe 1:** 140 g butter, 350 g of flour and 70 g of sugar.
Recipe 2: 110 g butter, 275 g flour and 55 g of sugar. The scones taste exactly the same. Can you use ratio to explain why?

In these next questions, think carefully about the information you have been asked for.

7 There are 8 dogs and 12 cats living in one street.

a) How many cats and dogs are there in total?

b) Express the ratio of dogs to the total number of cats and dogs in simplest form.

8 In an office there are 5 young people on apprenticeships and 50 experienced staff members. Find the ratio of the number of apprentices to the total number of apprentices and experienced staff. Express the ratio in its simplest form.

9 There are 60 teenagers, 32 adults and 4 children at a match.

a) Express the ratio of children to teenagers to adults in its simplest form.

b) Express the ratio of adults to children and teenagers in simplest form.

c) Express the ratio of teenagers to adults and children in simplest form.

In the following questions, remember to check that units are consistent.

10 Express each animal's birth to adult weight in its simplest form.

Animal	Birth	Adult
grey seal	15 kg	250 kg
capercaillie	80 g	1·84 kg
red deer	8000 g	152 kg
pine marten	120 g	1·68 kg
terrier	240 g	3 kg

11 Simplify:

a) 45 mins : 1 hour b) 12 mins : 1 hour c) 2 hours : 80 mins d) 48 mins : 3 hours

e) 180 cm : 3 m f) 7·2 m : 90 cm g) 20 cm : 2·5 m h) 15 mm : 2 m

i) 2 days : 2 weeks j) 5 weeks : 10 days k) 5 hours : 5 days l) 40 minutes : 1 week

12 In a school survey, 260 pupils brought packed lunches and 20% more used the cafeteria. Express the number of packed to cafeteria lunches as a ratio in its simplest form.

13 TV audiences are not happy when the ratio of adverts to tv shows is more than 1 : 50. During a show that lasted a full 25 minutes, 90 seconds of adverts was shown in between. Do you think the audience was happy? Explain clearly.

▶ Using ratio

When we know a ratio and one amount we can calculate the related quantity or quantities.

- Set down appropriate headings, the basic ratio and the known quantity.
- Calculate the multiplying factor.
- Use the multiplying factor to scale the ratio. Multiply each part of the ratio by the same amount.

Worked examples

1 A nursery has an adult to child ratio of $2:7$.

   How many adults are needed to look after 28 children?

   adults : children

   $28 \div 7 = 4$

   $\times 4$ ($2:7$) $\times 4$

   multiplying factor is 4 $8:28$ 8 adults are needed.

2 The ratio of flats, terraced houses and villas on a new estate is $5:2:1$. There are 38 terraced houses being built. Calculate the total number of new homes on the estate.

   $\dfrac{19}{2\overline{)3^{1}8}}$

   flats : terraced : villas

   $5 : 2 : 1$

   The multiplying factor is 19. $\times 19$

   $95 : 38 : 19$

   $\begin{array}{r} 19 \\ \times\ 45 \\ \hline 95 \end{array}$

   95 flats, 38 terraced houses, 19 villas

   $95 + 38 + 19 = 152$ new homes.

Exercise 4B

1 Copy each ratio and calculate the unknown quantities for each one.

   a) $2:3$ b) $5:7$ c) $8:3$ d) $9:5$
       $12:$ $35:$ $:27$ $:60$

   e) $3:4$ f) $2:3$ g) $7:2$ h) $7:11$
       $90:$ $150:$ $:80$ $:121$

2 Copy and complete:

   a) adult : child   b) red : white   c) houses : flats   d) staff : guests
       $2:7$ $3:5$ $4:9$ $8:150$
       $:42$ $42:$ $56:$ $64:$

3 On a school trip the ratio of teachers to pupils is $1:9$. How many pupils can 5 teachers take on the trip?

4 On another trip the ratio of teachers to pupils is $1:13$. How many teachers do 143 pupils need? ➡

5   Apple and pear purees are mixed in the ratio 3:2. How much pear puree mixes with 240 ml of apple?

6   Juice is diluted with water in the ratio 2:9, two parts juice to nine parts water.
    a)   How much water is needed for 50 ml of juice?
    b)   How much juice is mixed with 405 ml of water?

7   A salon apprentice is given these instructions for mixing hair dye and treatment.

    > dye : treatment
    > 1:1·5

    a)   Show that the ratio of dye to treatment can also be written as 2:3.
    b)   How much treatment do they need to mix with 440 ml of dye?
    c)   How much dye do they need to mix with 135 ml of treatment?

8   A construction apprentice is given these instructions for making mortar.

    > cement : sand
    > 3:4

    a)   How much sand do they need to mix with 1500 g of cement?
    b)   How much cement do they need to mix with 1624 g of sand?

9   The ratio of senior citizens to adults to teenagers on a bus is 5:19:7. There are 15 senior citizens on the bus. Calculate:
    a)   the number of adults and the number of teenagers
    b)   the total number of people on the bus.

10   A pupil is building a model of Aspirin $C_9H_8O_4$; one molecule has carbon, hydrogen and oxygen atoms in the ratio 9:8:4. The pupil uses 36 model carbon atoms. Calculate the total number of 'atoms' in the completed model.

11   A charity donates to schools and hospitals in the ratio 20:7. Schools received £4000. Find the total amount donated.

12   An online shop has a sales to returns ratio of 25:7. They received 133 returns in January. Calculate the total number of transactions (sales + returns) made.

13   Libbie splits her time between site visits, meetings and her office in the ratio 2:5:3. In one month, she completed 65 hours of meetings. Calculate the total number of hours Libbie worked that month.

14   A special porridge mix is made using oats and dried fruit in the ratio 9:1. The oats cost £2 per kg which is 25% of the price of the fruit. Calculate the total cost of a mixture which contains 500 g of dried fruit. Assume you can buy only as much as you need.

15   A milkshake recipe uses fruit puree and milk in the ratio 1:4. Megan has 2 litres of milk and 300 ml of fruit puree.
    a)   Can Megan make enough smoothie to use up all the milk? Explain your answer clearly.
    b)   What is the maximum volume of milkshake she can make?

16   A manufacturer is making brass using copper and zinc in the ratio is 12:7. We will assume there is no loss of mass when the materials are combined. The manufacturer has exactly 240 kg of copper and 133 kg of zinc.
    a)   Can the manufacturer make brass which uses all of the copper? Explain your answer clearly.
    b)   What is the maximum weight of brass which can be made?

17   Two drinks are made using cordial and water. The first uses a ratio of 2:3 with 150 ml of cordial. The second uses a ratio of 1:4 with 500 ml of water. The drinks are then mixed together. Find the ratio of cordial to water in the new drink.

▶ Sharing in a given ratio

We can share a total quantity into amounts according to a given ratio.

- Add the parts of the ratio.
- Divide the total by the sum of the parts to find the value of one share.
- Multiply each part by the value of one share.

Worked examples

1   A man wants to share his savings between two charities, Wildlife UK and Animal Rescue. He has £4000 to donate and wants to share the money in the ratio 3:2.

Wildlife UK : Animal Rescue

$3 + 2 = 5$ $5\overline{)4^4000}$ $\begin{array}{r} 800 \\ \times\ 3 \\ \hline 2400 \end{array}$ $\begin{array}{r} 800 \\ \times\ 2 \\ \hline 1600 \end{array}$

3 : 2
×800 ⤾ £2400 : £1600 ⤿ ×800

Check 2400 + 1600 = £4000

2   White, yellow and blue paint is mixed in the ratio 4:3:1. The mixed paint is green and is sold in 3-litre tins. How much white, yellow and blue paint is needed to make one tin of green?

white : yellow : blue

$4 + 3 + 1 = 8$ $8\overline{)3^30^60^40}$ $\begin{array}{r} 375 \\ \times_{32}4 \\ \hline 1500 \end{array}$ $\begin{array}{r} 375 \\ \times_{21}3 \\ \hline 1125 \end{array}$ $\begin{array}{r} 375 \\ \times\ 1 \\ \hline 375 \end{array}$

4 :   3 :   1   ×375
1500 ml : 1125 ml : 375 ml

Check 1500 + 1125 + 375 = 3000 ml

Exercise 4C

1   Share each amount in the given ratio.

   a)   50 in the ratio 2:3   b)   54 in the ratio 2:7   c)   63 in the ratio 5:2

   d)   720 in the ratio 4:5   e)   270 in the ratio 7:2   f)   121 in the ratio 6:5

   g)   260 in the ratio 7:3   h)   1080 in the ratio 5:1

2   Two friends share £49 in the ratio 2:5. How much will each receive?

3   Two friends split a 280 g treat bar in the ratio 3:4. How much did each receive?

4   An athlete splits her time into training and rest days in the ratio 3:2. Calculate how many training days she will have over a 365-day year.

5   Two farmers split 108 kg of animal feed in the ratio 7:5. How much feed will each farmer have?

6   An oil worker works for 13 days in a row and then has 5 days off. Calculate how many working and holiday days she will have over a contract lasting 180 days.

7   A business woman keeps and reinvests her profits in the ratio 4:3. Her company made a total profit of £47250 last year. Calculate the amount of money which was reinvested.

➡

8 Share each amount in the given ratios. Begin by adding all three parts of the ratio.

   a) 121 m in the ratio 2:4:5
   b) 360 g in the ratio 3:1:4
   c) £24 in the ratio 9:1:10
   d) 8 kg in the ratio 5:3:2
   e) 280 mm in the ratio 7:2:5
   f) 1 day in the ratio 4:3:1
   g) 1 hour in the ratio 8:3:1
   h) 3 hours in the ratio 4:3:8

9 Share 90 m² of shop floor space amongst shoes, bags and hats in the ratio 9:4:2. How much space will each have?

10 A scone mix weighs 1720 g and contains flour, butter and sugar in the ratio 5:2:1. Calculate how much of each ingredient the mix contains.

11 A manufacturer sells to Scotland, Wales and England in the ratio 19:1:20. The total value of sales one year was £216000. Calculate the value of sales in each part of the UK.

12 The table shows the basic ratios used to mix different colours of paint.

   a) How much red paint is needed to make 1·5 litres of Vermillion?
   b) How much red paint is needed to make 4·5 litres of Magenta?
   c) How much blue paint is needed to make 1·82 litres of Green?
   d) How much yellow paint is in 2·94 litres of Green?

Vermillion	red : yellow
	2 : 1
Magenta	red : blue
	2 : 1
Green	red : blue : yellow
	1 : 4 : 2

In these questions think carefully about the information you have been asked for.

13 In a school with 570 pupils the ratio of junior to senior pupils is 10:9. How many more juniors than seniors are there?

14 A building brick castle contains 1440 bricks. The bricks are grey, black and white in the ratio 4:3:5. Calculate the total number of grey and black bricks used.

15 A shopping centre has a total of 800 parking spaces. The ratio of standard, disability and family spaces is 15:3:2. How many more standard than family spaces are there?

16 The tables below show the basic ratios used to make different kinds of coffee and the sizes available.

	Coffee	Milk	Water	Foam
Americano	1	0	4 hot	0
Cappuccino	2	1 hot	0	2
Latte	2	3 hot	0	1
Frappe	1	2 cold	2 cold	1

Small	180 ml
Regular	330 ml
Large	450 ml

   a) Calculate the volume of coffee and water in a small Americano.
   b) Write out the recipe for a large Frappe. Calculate the volume of each ingredient.
   c) How much coffee is needed for an order of 3 regular Lattes?
   d) What contains more foam, a regular Cappuccino or a large Frappe?
   e) Which regular drink will contain the most coffee: Americano, Cappuccino, Latte or Frappe? Explain your answer.

Worked example

The ratio of cafeteria to packed lunches in a primary school is 2:1. What fraction of the pupils use the cafeteria?

Cafeteria   Packed

$\dfrac{2}{3}$ use the cafeteria

- The numerator is the relevant part of the ratio.
- The denominator is the sum of the parts of the ratio.

17 The ratio of passes to fails at a driving test centre one month was 9:3. What fraction of people passed their test?

18 An air steward spends time at home and abroad in the ratio 2:7. What fraction of his time does he spend at home?

19 A copper rod is divided in the ratio 3:2. The larger part is used for plumbing. What fraction is left?

20 A gardening group shares their plot to grow carrots, potatoes and turnips in the ratio 1:6:1.

   a) What fraction of the land is used to grow potatoes?

   b) What percentage of the land is used to grow carrots?

21 Look again at the table of basic ratios used to make different kinds of coffee.

	Coffee	Milk	Water	Foam
Americano	1	0	4 hot	0
Cappuccino	2	1 hot	0	2
Latte	2	3 hot	0	1
Frappe	1	2 cold	2 cold	1

   a) What fraction of an Americano is coffee?

   b) What percentage of a Cappuccino is foam?

   c) Think again about which drink will contain the most coffee. Order the drinks from strongest to weakest using fractions or percentages to explain your answer.

22 A primary school teacher has three groups in his class. He shares a total of 248 gold stars amongst them one week. He works out the ratio in which the stars were shared and notices: all three parts of the ratio are unique prime numbers and the sum of the parts is also prime. Determine two different ways the stars could have been shared.

23 A mother shares £1655 amongst her three daughters. She gives the middle daughter 10% more than the youngest and the oldest 10% more than the middle daughter. How much money did each daughter receive?

▶ Direct proportion

When quantities are in **direct proportion** they scale down and up identically. If one goes down, so does the other and if one goes up, so does the other.

- Set the information down under appropriate headings.
- Scale down by dividing each side by the same amount.
- Scale back up by multiplying each side by the same amount.

In this section, apart from Q10, assume the quantities described are in direct proportion.

Worked examples

1 An apprentice engineer works for 4 hours and is paid £36.
How much will they be paid for 7 hours work?

Time (h)	Wage (£)
4	36
1	9
7	63

$\div 4$ ... $\div 4$ ... $\times 7$ ... $\times 7$

They will be paid £63 for 7 hours.

2 A machinist makes 9 t-shirts in 45 minutes.
How long will it take to make 12 shirts?

Number of t-shirts	Time (minutes)
9	45
3	15
12	60

$\div 3$ ... $\div 3$ ... $\times 4$ ... $\times 4$

It will take 1 hour to make 12 t-shirts.

Exercise 4D

1 Set down and complete these direct proportion problems.

a)
Number of carrots	Portions
10	20
1	
8	

b)
Time (h)	Wage (£)
8	56
1	
12	

c)
Number of guests	Cost (£)
15	9
5	
40	

d)
Number of People	Rice (g)
8	200
12	

e)
Pages	Print Time (seconds)
42	70
6	
30	

f)
Portions	Cost (£)
16	24
20	

2 To make porridge for 3 people a cook uses 120 g of oats. What weight of oats is needed for 8 people?

3 It takes a team of forest workers 2 days to plant 100 trees. How many trees will they plant in 7 days?

4 A 5-night stay in the hotel costs £325. What will a 9-night stay at the same rate cost?

5 A large office fills 20 recycling bins in 5 days. How many bins will they fill in 40 days?

6 It takes a shop 12 minutes to serve 8 customers. How long will it take to serve 20 customers?

7 In 12 minutes a canteen worker can serve 30 people. How many people can he serve in 16 minutes?

8 A printer takes 18 minutes to produce 360 leaflets. How many leaflets will be printed in 45 minutes?

9 In 20 minutes a bakery fan removes 110 m³ of air. What volume of air will be removed in 90 minutes?

10 Are the following quantities in direct proportion? Show working to explain your answer.

   a) A 3 pack of magazines costs £8·52. A 5 pack of magazines costs £14·20.

   b) A course of 8 driving lessons costs £204. A course of 12 lessons costs £300.

   c) In 5 minutes, a cup of tea cools down by 10 °C. In a further 15 minutes it cools by 20 °C.

   d) A colony of 30 penguins eats 57 kg of fish a day. A colony of 50 penguins eats 95 kg of fish a day.

   e) With 40 minutes of charge a battery lasts 2 days. With 90 minutes of charge the battery lasts 6 days.

11 At an event with 1200 people in the audience there are 16 people checking tickets. How many people are needed to check tickets at an event with 3000 people in the audience?

12 In 8 minutes a cable car rises 120 m. How long will the cable car take to complete a 540 m climb?

13 The table shows the results of a survey. Three people were asked how many hours they worked that day and how much they earned.

Job	Hours worked	Earnings (£)
Trainee	8	60
Technician	6	90
Coder	5	225

   a) How much would the trainee earn in 5 hours?

   b) Calculate how many hours the technician would have to work to earn the same as the coder did on this day.

   c) Copy and complete these tables of earnings.

Hours worked	1	2	3	4
Trainee earnings (£)				

Hours worked	1	2	3	4
Technician earnings (£)				

   d) Use the tables to draw a line graph of the Technician and Trainee's earnings on the same axes. Display time in hours on the x-axis and choose a scale for earnings (£) on the y-axis that will allow you to plot both lines clearly.

   e) What happens to the lines on your graph as the number of hours increases? Can you explain why?

14 At a wedding with 45 guests the hosts provided 180 canapes, 90 party favours and spent £1170 on main meals. If they invited 20 more guests:

   a) how many canapes and favours would they need?

   b) how much would the main meals cost?

15 A team of 3 robots paints 270 m² of metal in 20 minutes. What area could a team of 5 robots paint in 1 hour?

▶ Inverse proportion

When quantities are in inverse proportion they scale up and down inversely. This means if one increases, the other decreases.

- Set the information down under appropriate headings.
- Divide one side, multiply the other by the same amount.
- Now multiply one side and divide the other by the same amount. See the worked example below.

In this section, assume the quantities described are in inverse proportion.

Worked example

A team of 3 people takes 6 hours to build a wall. How long would the same job take 2 people?

Number of people	Time (h)
3	6
÷3	×3
1	18
×2	÷2
2	9

The product of the quantities stays the same.

There are 18 hours of work to do, no matter how it is shared.

The job would take 2 people 9 hours. With fewer people, the job takes longer.

Exercise 4E

1 It takes a team of 5 builders 4 days to complete the foundations of a house. How long would the same job take if 2 builders worked at the same rate?

2 A team of 3 teachers take 8 hours to mark exam papers. How long would 4 teachers take to do the same job?

3 A farmer has enough food to feed 20 cows for 30 days. How long would the food last 25 cows?

4 A raffle prize is shared equally amongst 5 winners and each person receives £560. Calculate the prize money if the raffle was changed so that 8 people shared the prize equally.

5 It takes a team of 12 people 50 minutes to set up for a concert. How long would the job take if 8 more people are hired?

6 Giulia has enough plant food to feed her 9 houseplants for 4 months. How long would the feed last if she only had 6 plants?

7 Alan has enough plaster to coat 6·5 m² of wall with plaster 2 mm thick. What area of wall will it cover if he makes the plaster 5 mm thick?

8 A bath tap which is 30% open fills a bath in 15 minutes. How long would the bath take to fill if the tap was fully open?

Check-up 👍

1 Express each ratio in its simplest form:

a) 21:28   b) 16:60   c) 48:144

d) 35:100:20   e) 490:140:700   f) 560:800:160

2 A car park has 42 white and 28 grey cars parked in it. Express the ratio of white to grey cars in its simplest form.

3 A playlist has 38 rock songs and 95 pop. Express the ratio of rock to pop songs in its simplest form.

4 A taxi driver completes 224 short and 84 long journeys. Express the ratio of short to long journeys in its simplest form.

5 A newly hatched eaglet weighs 900 g. An adult eagle weighs 3·6 kg. Express the chick to adult weight in its simplest form.

6 A sports drink is made using 250 ml of concentrate and 2 litres of water. Express the ratio of concentrate to water in its simplest form.

7 The ratio of staff to guests in a hotel is 2:15. How many guests can 20 staff accommodate?

8 The ratio of nuts to fruit in a snack mix is 5:4.

a) A small pack contains 60 g of nuts. Calculate the weight of fruit in a small pack.

b) A large pack has 88 g of fruit. Calculate the weight of nuts in a large pack.

9 A hotel displays red and white roses in the ratio 2:3.

a) On the reception desk, there are 28 red roses. How many white roses should there be?

b) The dining room has 84 white roses. How many red roses should there be?

10 A smoothie recipe mixes mango, apple and pear in the ratio 2:5:6. A large smoothie has 180 ml of pear juice. Calculate the volumes of mango and apple juice in a large smoothie.

11 Over a year a business keeps the value of its stock, sales and savings in the ratio 3:8:5.

a) The business made sales of £64000 one year. Calculate the value of its stock and savings that year.

b) Calculate the total value (the sum of the stock, sales and savings) of the business that year.

12 Share each total in the given ratio:

a) 56 m in the ratio 5:2   b) 900 g in the ratio 5:1

c) £350 in the ratio 2:3   d) £45 in the ratio 5:3:2

13 A chef is making 2800 g of fruit compote using apricots and strawberries in the ratio 5:3. Calculate the weight of apricots and strawberries they will use.

14 A recycling plant receives plastics, paper and tin in the ratio 4:2:1. How much of each will be recycled in a week when the plant receives 560 tonnes of waste.

15 A GP sees patients in her surgery and at home in the ratio 13:2. One week she saw 10 patients at home. How many more patients did she see in the surgery?

16 A company shares £72000 amongst three projects in the ratio 17:3:10. Calculate each amount allocated.

→

17 Spot the odd one out in each list:

	List 1
A	28:56
B	19:38
C	16:36

	List 2
A	72:45:18
B	240:50:80
C	400:250:100

	List 3
A	20:55:15
B	48:60:36
C	36:45:27

18 In a school the ratio of pupils walking to school and taking the bus is 7:1. What fraction of pupils take the bus?

19 An engineer works for 19 days offshore, 3 days in Aberdeen and 2 days in Edinburgh. What fraction of her time does she spend in Aberdeen?

20 a) A gym has 42 treadmills, 35 rowing machines and 14 exercise bikes. Express the ratio of treadmills to rowing machines to bikes in its simplest form.

   b) The gym has full and swimming-only memberships in the ratio 10:3. There are 840 swimming-only members. Calculate the number of full members.

   c) The gym runs cardio, martial arts and strength classes in the ratio 3:1:2. There are a total of 78 classes every week. Calculate the number of each kind of class that are on each week.

In questions 21–24 assume that the quantities are in direct proportion.

21 In 7 days a restaurant kitchen uses 280 litres of stock. Calculate how much stock the kitchen will use in 25 days.

22 A car can travel 99 miles with 9 litres of petrol. Calculate the distance the car can travel with 40 litres of petrol.

23 A worker checks 20 circuit boards in 15 minutes. How many will she check in 24 minutes?

24 On a day with 1200 visitors a park ranger collects 16 bags of rubbish. How many bags of rubbish would you expect on a day with 2100 visitors?

In questions 25 and 26 assume that the quantities are in inverse proportion.

25 It takes a team of 6 roadies 4 hours to set up the stage and lighting for a concert. How long would the same job take if 8 roadies worked on it?

26 A troop of 18 soldiers has enough rations to last 5 days. How long would the same rations last a troop of 15 soldiers?

27 A school enterprise group is planning a fund-raising day. The group consists of 4 teachers, 9 senior pupils and 7 junior pupils.

   a) The junior pupils plan to raise £385. The seniors hope to raise twice as much per pupil. Calculate the total amount the seniors hope to raise.

   b) What fraction of the group are teachers?

   c) A float of £100 in change is to be shared amongst all the volunteers at the event. Calculate the value of the float which the junior pupils will receive.

   d) Last year the teachers set up the event on their own, taking 25 minutes to do so. If all the pupils help and everyone works at the same rate, how long will it take to set up the event?

5 Time

▶ Time intervals

Time does not follow the decimal system so you cannot simply add or subtract times to calculate a time interval. Instead, you can use a timeline like this one.

When you draw a timeline, make sure you label all the times clearly.

- Count on to the next/required hour.
- Count on to the required minutes.
- Add all time intervals together to form your answer.

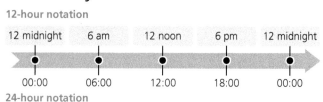

Note: Sometimes it is easier and quicker to count on to the hour past the required time and then count back by the appropriate number of minutes.

Worked examples

1 Calculate the length of each time interval, in hours and minutes.

   a) 8 am to 5:15 pm.

   b) 20:45 to 00:20.

2 A train sets out at 9:30 am and reaches its destination at 2:18 pm. How long does the journey take?
   Give your answer in hours and minutes.

3 A new bike tyre is fitted at 16:40 on 9 June. The tyre is punctured at 14:15 on 24 June. How long was the tyre on the bike before it was punctured?

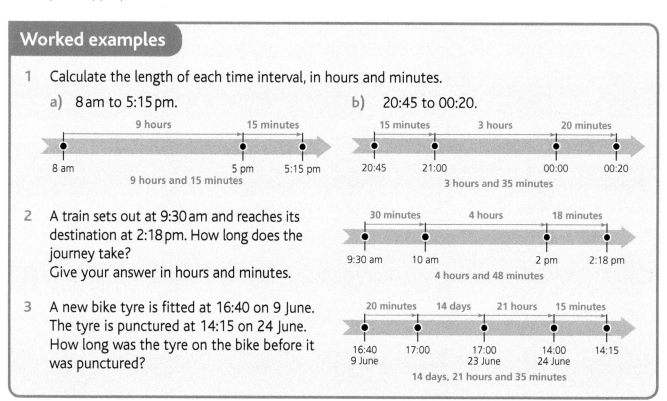

Exercise 5A

1 Calculate the time interval, in hours and minutes, from:

   a) 7:30 am to 11:45 am

   b) 3:40 pm to 7:15 pm

   c) 10:45 pm to 6:50 am

   d) 3:16 am to 1:05 am.

2 A school day starts at 8:45 am and finishes at 3:35 pm. How long, in hours and minutes, is the school day?

3 How long is it between these times? Give your answers in hours and minutes.

a) 05:40 and 11:52   b) 14:20 and 19:08

c) 09:10 and 16:14   d) 17:55 and 09:20

4 A roasting joint was put in the oven at 13:20 and removed at 15:50.

a) How long was the joint in the oven for? Give your answer in hours and minutes.

Cooking instructions state the joint should cook for half an hour per 500 g plus an extra half an hour.

b) If the instructions were followed correctly, what weight, in kilograms, was the joint?

5 Calculate these time intervals:

a) 6 pm on 17 February until 9 pm on 27 February

b) 2:45 pm on 5 August until 1 pm on 25 August.

6 This is part of a bus timetable.

The journey times between stops for each bus are the same as for the first one.

a) Copy and complete the timetable.

b) How long does a bus take to travel from Sandbank to Millhouse?

c) Aisha was on the bus for 36 minutes. Between which two stops did she travel?

Location	Bus 1	Bus 2	Bus 3	Bus 4
Dunoon	06:27	07:30		
Sandbank	06:35		07:58	
Tighnabruaich	07:03			
Kames	07:21	08:24		
Millhouse	07:24	08:27	08:47	
Portavadie	07:32			10:05

d) Morag lives in Tighnabruaich. It takes her 15 minutes to walk from the bus stop in Portavadie to her work. If she starts work at 9 am, what is the latest time she can get a bus from Tighnabruich to make sure she gets to work on time?

7 Mariah is travelling from Glasgow to London. She obtains this information.

Train	Departs	Arrives	Cost
Sleeper	20:10	07:01	£119·50
Express	19:45	00:25	£140

a) How much quicker is the Express than the Sleeper train? Give your answer in hours and minutes.

b) What is the cost per minute of the Express train?

c) The Express train is delayed by 52 minutes. What time will it arrive in London?

8 A sailing instruction course between Largs and Oban is taking place during 6–11 May. It runs from 5 pm on the first day until 3:30 pm on the final day. How long does the course last?

9 A major golf tournament starts at 07:00 on 16 September. The winner putts the final shot at 15:44 on 19 September. How long was it from the start of the tournament to the winning putt?

10 To raise money for charity, John cycles from John O'Groats to Land's End. He leaves John O'Groats at 10:30 am on 7 March and arrives at Land's End at 09:19 am on 18 March. How long did his journey take?

11 A plane journey to Australia takes 27 hours. The Millar family left Scotland for Australia on 14 July at 2 pm GMT (UK time).

a) What was the date and time in the UK when the Millars landed in Australia?

The time in Australia is 9 hours ahead of UK time.

b) What was the date and time in Australia when the Millars landed?

▶ Calculating speed

It is natural to speed up and slow down over the course of a journey. For the remainder of this chapter we will refer to the average speed of a journey, not the speed at any one given time.

We can calculate the speed of any journey from the formula speed = $\dfrac{\text{distance}}{\text{time}}$ or $S = \dfrac{D}{T}$.

To calculate a speed:

- Write the formula down.
- Substitute appropriate values into the formula (take care with units!).
- Calculate the answer and include units.

Worked examples

1 A car travels 105 miles in 2 hours. What speed is the car travelling at?

$$S = \frac{D}{T}$$
$$= \frac{105}{2}$$
$$= 52{\cdot}5\,\text{mph}$$

2 A plane takes 30 hours to fly 18 000 km. What speed is the plane flying at?

$$S = \frac{D}{T}$$
$$= \frac{18000}{30}$$
$$= 600\,\text{km/h}$$

3 Alex runs 2·4 kilometres in 10 minutes. Calculate Alex's average speed, in metres per second.

Rewrite units to be consistent with the required units in the answer.

2·4 kilometres = 2400 m and 10 minutes = 600 seconds.

$$S = \frac{D}{T}$$
$$= \frac{2400}{600}$$
$$= 4\,\text{m/s}$$

Exercise 5B

1 Calculate the average speed for each journey. Be sure to include units with your answer.

	a)	b)	c)	d)	e)	f)	g)
Distance	140 miles	52 metres	1500 km	360 miles	950 metres	13252 km	480 metres
Time	7 hours	4 seconds	6 hours	5 hours	2 seconds	4 hours	20 seconds

2 A border collie dog ran 130 metres in 10 seconds. Calculate the dog's average speed.

3 Abdullah drove 162 miles from Kelso to Campbeltown. He left his house at 10:15 and arrived at 14:15. What was his average speed?

4 Emma won a swimming competition. She swam 150 metres in 1 minute and 40 seconds. What was her average speed, in metres per second?

5 Mr Clough is a dog walker. He believes his average step is 60 cm. If he walks 10 000 steps in 2 hours, what speed did he walk at? Give your answer in km/h.

6 Atif ran 5·1 miles in half an hour. George cycled 2·2 miles in 10 minutes. How much faster was George than Atif? Give your answer in a) miles per minute   b) miles per hour.

▶ Calculating distance

We can calculate the distance of any journey using the formula distance = speed × time, which we can write as $D = S \times T$.

Worked examples

1 Priyanka cycled for 2 hours at a speed of 8 mph.
   How far did she cycle?
   $D = S \times T$
      $= 8 \times 2$
      $= 16$ miles

2 Jack drives at an average speed of 45 mph for 3 hours. How far does he drive?
   $D = S \times T$
      $= 45 \times 3$
      $= 135$ miles

3 Alison crosses a park on a Segway. It takes her one and a half minutes travelling at an average speed of 5 metres per second. What distance does she travel?
   $D = S \times T$ Time has to be in seconds to be consistent with the unit for speed.
      $= 5 \times 90$
                    One and a half minutes = 90 seconds.
      $= 450$ metres

Exercise 5C

1 Calculate the distance travelled for each journey. Be sure to include units with your answer.

	a)	b)	c)	d)	e)	f)	g)
Speed	60 mph	2 m/s	300 km/h	35 mph	23 m/s	143 km/h	2300 km/h
Time	3 hours	50 seconds	4 hours	8 hours	40 seconds	20 hours	60 hours

2 Kurt walks at an average speed of 1·5 metres per second. How far can he walk in 40 seconds?

3 A plane flew for 3 hours at an average speed of 550 mph. How far did it fly?

4 Jasmine left home at 23:15 to drive to the airport for an early morning flight. She drove at 42 mph and arrived at the airport at 02:15. What distance did she drive?

5 A cheetah can sprint at 28 metres per second. How far will it cover in 2 minutes? Give your answer in kilometres.

6 Euan and Thomas work on a farm. Euan took a quad bike to check on sheep at the top of a hill. He took 2 hours and drove at an average speed of 20 mph.

   a) Calculate the distance travelled by Euan.

   Thomas took a tractor to the local market. His average speed was 33 km per hour and he also took 2 hours.

   b) How far did Thomas travel?

   One mile is approximately 1·6 km.

   c) Who travelled further and by how far? Give your answer in metres.

7 The average speed of a common snail is 1 mm per second.

   How far could a snail travel in 1 day? Give your answer in metres.

▶ Calculating time

We can calculate the time taken for any journey from the formula time $= \dfrac{\text{distance}}{\text{speed}}$, which we can write as $T = \dfrac{D}{S}$.

You should already be familiar with these time conversions.

Time	Decimal hour	Minutes
Half an hour	0·5	30
Quarter of an hour	0·25	15
Three quarters of an hour	0·75	45

Worked examples

1   A train travels 475 km at a speed of 100 km/h. How long does the train journey take?

$$T = \frac{D}{S}$$
$$= \frac{475}{100}$$
$$= 4·75 \text{ hours} = 4 \text{ hours and } 45 \text{ minutes}$$

2   A snake slithers at 8 mph.

How long did the snake take to slither 12 miles?

$$T = \frac{D}{S}$$
$$= \frac{12}{8}$$
$$= 1·5 \text{ hours} = 1 \text{ hour and } 30 \text{ minutes}$$

3   Qasim competes in open water swimming competitions. In his last competition, his average speed was 2·5 km/h. How long will it take Qasim to complete a 10000 metre race if he can maintain the same average speed?

Distance has to be in kilometres to be consistent with the unit for speed.

10 000 metres = 10 km

$$T = \frac{D}{S}$$
$$= \frac{10}{2·5}$$
$$= 4 \text{ hours}$$

Exercise 5D

1   Calculate the time taken for each journey. Be sure to include units with your answer. Convert any answers that include a decimal part of an hour to hours and minutes.

	a)	b)	c)	d)	e)	f)	g)
Distance	250 miles	78 metres	420 km	1170 m	8778 km	370 miles	4175 km
Speed	10 mph	6 m/s	5 km/h	9 m/s	7 km/h	40 mph	50 km/h

2   Hannah has a hoverboard. She completed a 35-metre obstacle course at an average speed of 5 m/s. How long did it take Hannah to complete the course?

3   A satellite with a high earth orbit travels at 10000 mph.

Calculate the time taken for the satellite to travel 5000 miles. Give your answer in minutes.

4   A delivery driver has to make two deliveries. She starts in Stirling and drives 120 miles to Carlisle at 40 mph. She then stays in Carlisle for 45 minutes before driving 125 miles to Port Glasgow at 50 mph. If the driver left Stirling at 10:50 am, when did she reach Port Glasgow?

5   John cycles at 8 m/s. How long will it take him to cycle 1·8 kilometres? Give your answer in minutes and seconds.

► A mixture!

You may find this memory aid helpful.

To find the correct formula, cover the letter you need to calculate. If you are asked to calculate the distance, for example, cover the 'D' with your finger to leave S × T.

Remember to:

● Show all working.

● Make sure units are consistent and include them in your final answer.

Exercise 5E

1 Calculate the missing entry for each journey.

	a)	b)	c)	d)	e)	f)	g)
Distance	60 metres		360 km	728 km		1470 km	30 km
Speed		14 mph	90 km/h		2·8 m/s		12 km/h
Time	6 seconds	3 hours		8 hours	40 seconds	5 hours	

2 The speed limit on a motorway is 70 mph. Caleb has 3 hours to drive 189 miles.

Can he make the journey without exceeding the speed limit? Justify your answer.

3 Louise rode her horse at an average speed of 7·5 mph. How far did she travel in 2 hours?

4 A small plane flies at an average speed of 200 km/h.

Calculate the time it takes to fly 650 km. Give your answer in hours and minutes.

5 At a fixed temperature, the speed of sound in air is 343 metres per second. Thunder can be heard 15 seconds after seeing the lightning strikes. How far away is the lightning? Give your answer in kilometres.

6 A train left the station at 17:00 and was due to arrive at its destination at 18:50.

Unfortunately, the train was 10 minutes late in arriving due to a fault on the line.

If the train travelled 184 000 metres, what was its average speed in kilometres per hour?

7 At football training, the team are told to run around the perimeter of the pitch three times.

James has an average speed of 4 m/s and completes the run in 2 minutes.

Calculate the length of the perimeter of the pitch.

8 Kitty competed in a triathlon. She

● swam 1500 metres at 20 metres per minute

● cycled 40 km at 16 km/h

● ran 10 km at 8 km per hour

● took 2 minutes to transition from swimming to cycling

● took 1 minute to transition from cycling to running.

a) If Kitty started the race at 11 am, when did she finish it?

b) She aims to compete in another triathlon in two months. She thinks she can reduce her swimming time to 50 minutes and maintain all her other times.

How much faster will she have to swim to achieve this?

▶ Distance–time graphs

● Look carefully at the scales on a distance–time graph to obtain values for distance and time.

Worked example

Robert leaves his house to give his friends Ellie and Murray a lift to their homes. He stops at both houses for a short time and Ellie is dropped off first.

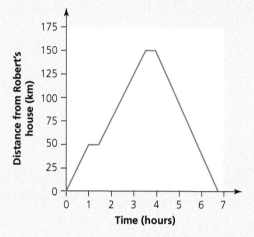

a) How far is it between Robert and Ellie's homes? — 50 km

b) How long did Robert spend at Ellie's house? — 30 minutes

c) What was Robert's average speed driving from Ellie's house to Murray's?

$$S = \frac{D}{T}$$
$$= \frac{100}{2}$$
$$= 50 \text{ km/h}$$

d) Where did Robert finish his journey? — He returned home as he was 0 km from where he started.

Exercise 5F

1   This distance–time graph represents a bus journey.

a) Calculate the average speed of the bus before its first stop.

b) How long did the bus stop for?

c) What was the average speed of the bus for the second part of the journey?

d) Choose the correct word to complete this sentence.

   The steeper the line the more quickly/slowly the bus is travelling.

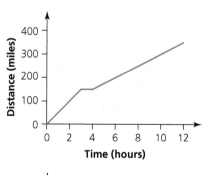

2   Heather and Taylor set out from the city centre to go to a football match.

a) Calculate their average speed when they were travelling to the stadium.

b) How long were they at the stadium for?

c) They walked straight home after the match was finished.

   What was their average walking speed?

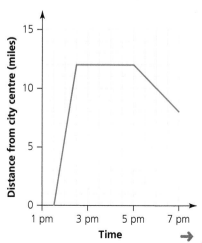

3   Tom works as a delivery driver. On one day, he had four deliveries to make and each delivery was made as soon as he stopped.

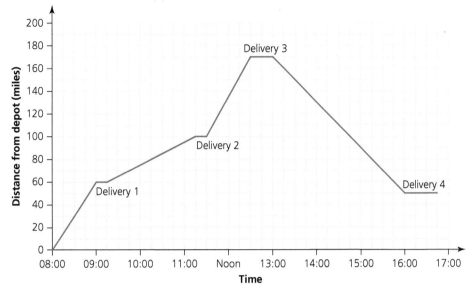

a) Calculate Tom's average speed on his way to each delivery.

b) He was asked to call his client half an hour before making delivery 2. At what time should he have called?

c) Tom had his lunch whilst making delivery 3. What was the maximum time he could have taken to eat his lunch?

d) Motorways and dual carriageways are the only roads in the UK where drivers can exceed 60 mph. Tom was only on the motorway once during the day and did not drive on a dual carriageway. Which delivery did he make after driving on the motorway?

e) After making delivery 4, Tom drives at 25 mph back to the depot. When did he arrive at the depot?

f) Tom gets paid for working 8 hours a day, excluding stops. He gets overtime for any time worked beyond this.

    How much overtime should Tom be paid for?

4   Alexa and Atif are training to take part in a fun run for charity. Alexa starts at the beginning of the route and jogs to the end, whilst Atif completes the same route but in the opposite direction.

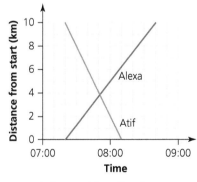

a) How long does it take Alexa to complete the route?

b) At what time do Alexa and Atif pass each other?

c) Calculate Alexa's average speed, in metres per minute.

d) How much greater was Atif's average speed than Alexa's? Give your answer in metres per minute.

5   Ms Cameron wanted to do some shopping. She left her home at 8 am and drove 40 miles in one hour. She stopped at a shopping centre for 2 hours before deciding to head home. After driving for half an hour she had travelled 15 miles and decided to stop for a coffee. She stayed in the coffee shop for 20 minutes. Once she finished her coffee, she took a further 40 minutes to get home.

    Illustrate Ms Cameron's journey on a distance–time graph.

6   Draw your own distance–time graph with at least one stop. Write a short story to accompany your graph, explaining clearly what is happening at each stage of the journey.

Check-up 👍

1. How long is it, in hours and minutes, from
   a) 04:15 until 12:32
   b) 9:40 pm until 12:51 am
   c) 5:44 am until 7:23 pm
   d) 10:32 until 01:18?

2. A film started at 2:50 pm and finished at 4:32 pm. How long did the film last?

3. Calculate the time interval from 6 am on 18 January until 12 noon on 21 January.

4. A landscape gardener started renovations on a garden at 08:30 on 5 April. She finished the job at 13:45 on 12 April. How long was it from the job starting to the garden being finished?

5. Calculate the speed if
   a) distance = 600 m, time = 10 seconds
   b) distance = 4000 km, time = 8 hours
   c) distance = 260 miles, time = 5 hours
   d) distance = 320 mm, time = 20 seconds

6. Calculate the distance if
   a) speed = 20 m/s, time = 9 seconds
   b) speed = 52 mph, time = 6 hours
   c) speed = 498 km/h, time = 20 hours
   d) speed = 60 cm per minute, time = 50 minutes

7. Calculate the time taken if
   a) distance = 200 metres, speed = 4 m/s
   b) distance = 42 miles, speed = 3 mph
   c) distance = 108 metres, speed = 6 m/s
   d) distance = 3800 km, speed = 40 km/h

8. A cruise ship sails at 25 mph. How far can it sail in 7 hours?

9. A hot-air balloon travelled 1225 km in 5 hours. What was the balloon's average speed?

10. Gill hit a target 30 metres away with a hockey ball.

    If the ball travelled at 20 m/s how long was it moving for before it hit the target?

11. An athlete leaves her house at 14:40 and returns home at 17:40. During this time, she cycles 48 miles. What was her average speed?

12. The graph shows David's motorbike journey.

    a) Calculate David's average speed prior to his first stop.

    b) How long was David's first stop?

    c) What was David's average speed for the second part of his journey?

    d) After his second stop, David continued until he reached his destination.

       During which part of David's journey did he travel the fastest?

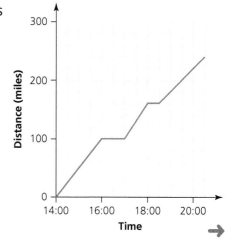

13 Sophie and Andrew are both travelling from Inverness to Aberdeen.

Sophie goes via Keith whilst Andrew makes the journey via Perth.

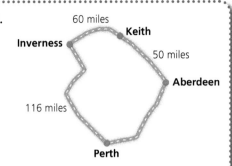

60 miles

Keith

Inverness

50 miles

Aberdeen

116 miles

Perth

   a) Sophie drove between Inverness and Keith at an average speed of 40 mph.

   How long, in hours and minutes, did it take her?

   b) There were roadworks between Keith and Aberdeen.

   Sophie left Inverness at 8 am and arrived in Aberdeen at 11:30 am.

   What was her average speed between Keith and Aberdeen?

   c) It took Andrew 2 hours to drive from Inverness to Perth.

   What was his average speed?

   d) Andrew's entire journey was half an hour longer than Sophie's.

   If he drove from Perth to Aberdeen at 42 mph, how much further did Andrew drive than Sophie?

14 Jake and Elena cycle the same course. Jake cycled for 4 hours at 15 mph.

Elena took 5 hours to complete the course. What speed did she cycle at?

15 A rabbit can run at 11 m/s. How far could it run in half a minute?

16 How much faster must you travel to cover 1 km in 40 seconds rather than 50 seconds?

Give your answer in metres per second.

17 A car travels at 90 km/h. How long will it take to travel 112 500 metres? Give your answer in hours and minutes.

18 The speed of light is approximately 300 000 km/s. How far, in kilometres, can light travel in one minute?

19 James drove 100 miles in 2 hours and Kara drove 210 km in 3 hours.

If one mile is approximately 1·6 km, how much faster, in km/h, did James drive than Kara?

20 A turtle can swim at 4 m/s. At a given temperature, the speed of sound under water is 1500 m/s.

   a) How many kilometres can the turtle swim in 20 minutes?

   b) How long will it take the turtle to hear a noise made 7·5 kilometres away?

21 A marathon is approximately 42 km. Fraz runs a marathon in 5 hours.

   a) Calculate Fraz's average speed in metres per minute.

   b) Fraz's ambition is to increase his speed to 200 metres per minute. How much faster will he complete his next marathon if he achieves this?

6 Measurement

▶ Units of measurement

Converting between metric units of length is an important skill. Try to memorise these facts.

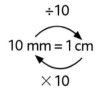
÷10
10 mm = 1 cm
×10

÷100
100 cm = 1 m
×100

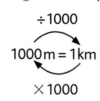
÷1000
1000 m = 1 km
×1000

To convert between metric units:

● Use the facts to decide which decimal calculation you need.

● Use place value to carry out the calculation.

Worked example   Write:

a) 85 mm in cm   85 mm   85 ÷ 10
                          = 8·5 cm

b) 59 cm in mm   59 cm   59 × 10
                          = 590 mm

c) 75 cm in m   75 cm   75 ÷ 100
                         = 0·75 m

d) 4·2 m in cm   4·2 m   4·2 × 100
                          = 420 cm

e) 4500 m in km   4500 m   4500 ÷ 1000
                            = 4·5 km

f) 0·85 km in m   0·85 km   0·85 × 1000
                             = 850 m

Exercise 6A

1   Convert these measurements to centimetres.
   a) 40 mm   b) 59 mm   c) 650 mm   d) 8 mm   e) 25 mm   f) 150 mm

2   Convert these measurements to metres.
   a) 240 cm   b) 126 cm   c) 520 cm   d) 3 cm   e) 86 cm   f) 4000 cm

3   Convert to kilometres.
   a) 3000 m   b) 2500 m   c) 750 m   d) 80 m   e) 7 m   f) 13700 m

4   Convert to metres.
   a) 6 km   b) 8·4 km   c) 12 km   d) 0·9 km   e) 4·071 km   f) 21·5 km

5   Convert to centimetres.
   a) 8 m   b) 3·1 m   c) 0·6 m   d) 15·02 m   e) 180 m   f) 0·02 m

6   Convert to millimetres.
   a) 5 cm   b) 4·2 cm   c) 0·5 cm   d) 24 cm   e) 54·2 cm   f) 179 cm

---

Worked example

Write 56580 cm in km.

56580 cm   One step at a time.
56580 ÷ 100
= 565·8 m   Write cm as m.
565·8 ÷ 1000
= 0·5658 km   Write m as km.

---

7   Rewrite these distances as indicated:
   a) 3480 mm in m   b) 14300 cm in km
   c) 903400 mm in m d) 7046 m in mm
   e) 7500 cm in km   f) 3 km in cm
   g) 8 m in mm   h) 1 mm in m

Weight and volume follow the same system.

$÷1000$ $÷1000$

$1000g = 1kg$ $1000ml = 1litre$

$× 1000$ $× 1000$

8   Convert to grams.
   a)   2 kg   b)   50 kg   c)   4·071 kg   d)   0·04 kg

9   Convert to kilograms.
   a)   8300 g   b)   1270 g   c)   650 g   d)   54 g

10   Convert to litres.
   a)   4000 ml   b)   920 ml   c)   51 ml   d)   7 ml

11   Convert to millilitres.
   a)   7·5 litres   b)   0·3 litres   c)   7·92 litres   d)   62·8 litres

>> Cubic units and liquid capacity

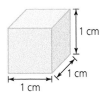

A hollow cubic centimetre holds exactly 1 millilitre of water.

$1 cm^3 = 1 ml$

$1000 cm^3 = 1000 ml = 1 litre$

Worked example

The volume of a hollow cuboid is $3750 cm^3$.

Calculate the **capacity** of the cuboid in litres.

$3750 cm^3 = 3750 ml$

$= 3750 ÷ 1000$

$= 3·75 litres$

12   These are the volumes of some hollow shapes. Calculate the capacity of each one in litres.
   a)   4500 cm^3 b)   7800 cm^3 c)   13 800 cm^3 d)   304 cm^3

>> Square units

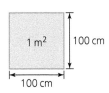

Square units do not follow the same pattern. This is a common mistake!

1 **square metre** is **not** the same as 100 **square centimetres**.

$1 m^2 ≠ 100 cm^2$

$1 m^2 = 100 cm × 100 cm$

$= 10000 cm^2$

$× 10000$

$1 m^2 = 10000 cm^2$

$÷ 10000$

13   Write each of these areas in square centimetres.
   a)   5 m^2 b)   2·9 m^2 c)   60 m^2 d)   0·25 m^2

14   Draw a centimetre square, label the sides in mm. Use the diagram to calculate how many square millimetres there are in a square centimetre. Write each of these in square millimetres.
   a)   9 cm^2 b)   65 cm^2 c)   0·8 cm^2

15   Draw a cube to represent 1 m^3 and label the sides in centimetres. Use the diagram to calculate how many cubic centimetres there are in a cubic metre.

   Write each of these in cubic centimetres:
   a)   4 m^3 b)   0·75 m^3 c)   58 m^3

▶ The area of a rectangle

The **area** of a shape is the amount of 2D space it occupies.

Area is measured in square units.

There are 3 rows of 4 square centimetres, $4 \times 3 = 12\,cm^2$.

To find the area of a rectangle or square:

Area = 1 cm²

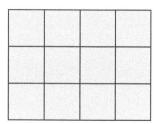

Area = 12 cm²

- Clearly state the formula: $A = l \times b$ or $A = lb$
- Substitute the correct values for the length l and breadth b
- Calculate and include square units

Area = length × breadth

$A = l \times b$

Worked examples

1 Calculate the area of this rectangle.

7 cm
12 cm

$A = l \times b$
$= 12 \times 7$ $l = 12$ and $b = 7$
$= 84\,cm^2$ units are cm^2

2 Calculate the area of this square.

5 mm

$A = l \times b$
$= 5 \times 5$ $l = b = 5$
$= 25\,mm^2$ units are mm^2

3 A rectangular swimming pool has the dimensions shown.

Calculate the area of the pool's base.

180 cm

5 m

$A = l \times b$ Units must be consistent.
$= 1.8 \times 5$ $l = 1.8\,m$, $b = 5\,m$, units are m^2.
$= 9\,m^2$

180 cm, 180 ÷ 100
$= 1.8\,m$

$\begin{array}{r} 1.8 \\ \times_4 5 \\ \hline 9.0 \end{array}$

Exercise 6B

1 Calculate the area of each rectangle or square.

a)

9 cm
5 cm

b)
7 cm
13 cm

c)
9 cm
9 cm

d)
45 m
20 m

2 Calculate the area of each rectangle.

a)
5.8 m
3 m

b)
6 cm
13.7 cm

c) Careful!
2 m
650 cm

d)
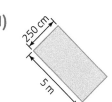
250 cm
5 m

3 A mirror ball is covered in 900 mirrored squares each with the dimensions shown. Calculate the total area of mirror on the ball.

7 mm

4 A gardener agrees to mow three lawns that have these dimensions.

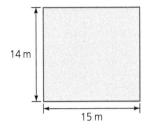

14 m

 5 m

6 m

6 m

6 m

20 m

15 m

a) Calculate the area of each lawn.

b) The gardener works at the same rate on every lawn. It takes 10 minutes to mow the first lawn. How long will it take to complete the rest of the job?

5 Notice the area of a square can be found using the formula $A = l \times l$ or $A = l^2$. Find the area of a square with sides of length

a) 7 cm
b) 40 mm
c) 11 cm
d) 15 m.

Worked example

The area of this rectangle is 84 cm².

Calculate the length of the rectangle.

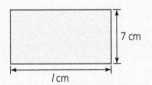

7 cm

l cm

$A = l \times b$

$84 = l \times 7$

$l = 12$ cm

Set down the formula.

Substitute all known values.

$$\frac{12}{7\overline{)8^14}}$$

---

6 Calculate the unknown dimension in each rectangle or square.

a)

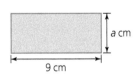

a cm

9 cm

Area = 27 cm²

b)

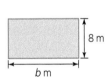

8 m

b m

Area = 96 m²

c)

40 mm

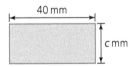

c mm

Area = 240 mm²

d)

d cm

d cm

Area = 900 cm²

7 A wall is painted in the colours shown.

a) Calculate the area to be painted blue.

b) Calculate the total area to be painted yellow.

c) Calculate the area to be painted white.

d) Calculate the area of the whole wall.

e) What percentage of the wall will be painted blue?

f) What fraction of the wall will be painted white?

g) Show that $\frac{5}{12}$ of the wall is painted yellow.

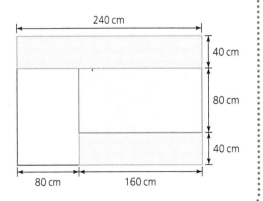

240 cm

40 cm

80 cm

40 cm

80 cm

160 cm

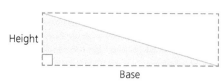

▶ The area of a triangle

A triangle occupies half of the area of the surrounding rectangle.

$$\text{Area} = \frac{1}{2} \times \text{base} \times \text{height}$$

$$A = \frac{1}{2} \times b \times h$$

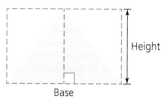

The height or altitude is perpendicular to the base. Perpendicular means 'at right-angles to'.

To find the area of a triangle:

- Clearly state the formula: $A = \frac{1}{2} \times b \times h$ or $A = \frac{1}{2}bh$.

- Substitute the correct lengths for the base b and height h.
- Calculate the area and include square units.

Worked examples

1 Calculate the area of the triangle.

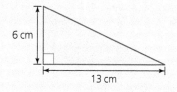

$$A = \frac{1}{2} \times b \times h$$

$$= \frac{1}{2} \times 13 \times 6$$

$$= 39\,\text{cm}^2$$

We can do the calculation in any order.

$$\begin{array}{r} 13 \\ \times\,6 \\ \hline 78 \end{array}\qquad \begin{array}{r} 39 \\ 2\overline{)7^18} \end{array}$$

or (easier when one of the numbers is even)
$6 \div 2 = 3 \quad 13 \times 3 = 39$

2 Calculate the area of the triangle.

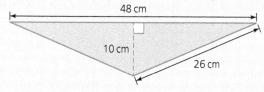

$$A = \frac{1}{2} \times b \times h$$

We **do not** use the sloping length.

$$= \frac{1}{2} \times 48 \times 10$$

The perpendicular height is 10 cm.

$$= 240\,\text{cm}^2$$

Exercise 6C

1 Calculate the area of each triangle. Be careful to use the correct square units.

a) 5 cm, 8 cm

b) 7 cm, 4 cm

c) 10 mm, 14 mm

d) 3 m, 5 m

2 Calculate the area of each triangle.

a) 5 mm, 7 mm

b) 7 m, 9 m

c) 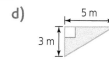 57 mm, 40 mm

d) 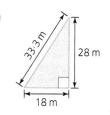 33.3 m, 28 m, 18 m

3 A jeweller makes a pair of identical stud earrings. Each earring is in the shape of a triangle. Calculate the total area of the visible triangular surfaces of the earrings.

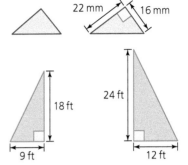

4 A sailing boat has two triangular sails with the dimensions shown. Calculate the difference between the areas of the larger and smaller sails.

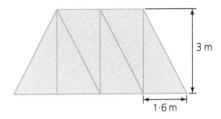

5 Structural engineers use triangles to add strength to structures, these are called trusses. The diagrams show trusses which are made from congruent (identical) triangles. Calculate the total area of each truss.

a)

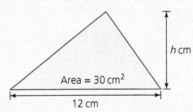

b)

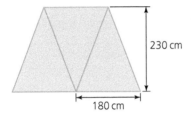

Worked example

Calculate the height of the triangle.

$A = \dfrac{1}{2}bh$ Write down the formula.

$30 = \dfrac{1}{2} \times 12 \times h$ Substitute all the known values, $A = 30, b = 12$

$60 = 12 \times h$

$h = 5 \text{ cm}$

6 The area of each triangle is given. Calculate the unknown dimension.

a) Area = 20 cm²

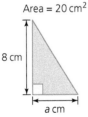

b) Area = 28 mm²

c)

d) Area = 112 cm²

7 In an obtuse-angled triangle the perpendicular height falls outside the shape.

a) Calculate the area of the shaded triangle. Use $A = \dfrac{1}{2}bh$.

b) Calculate the area of the shaded triangle by subtracting the area of one right-angled triangle from another.

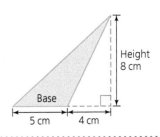

▶ The area of a parallelogram

A **parallelogram** occupies an area which is equal to the area of a rectangle with the same dimensions.

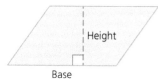

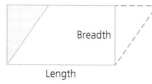

Area = base × height
$A = b \times h$

To find the area of a parallelogram:

● Clearly state the formula: $A = b \times h$ or $A = bh$.

● Substitute the correct values for the base b and perpendicular height h.

● Calculate and include square units.

Worked example Calculate the area of each parallelogram.

a)

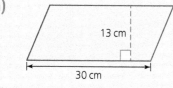

$A = b \times h$
$= 30 \times 13$
$= 390 \, cm^2$

$13 \times 3 = 39$
$39 \times 10 = 390$

b)

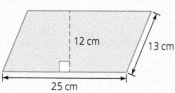

$A = b \times h$
$= 25 \times 12$
$= 300 \, cm^2$

We do not use the sloping length; the height is perpendicular to the base.

c)

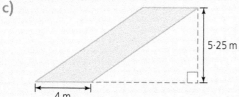

$A = b \times h$
$= 4 \times 5{\cdot}25$
$= 21 \, m^2$

$\begin{array}{r} 5{\cdot}25 \\ \times \quad _1{}_24 \\ \hline 21{\cdot}00 \end{array}$

The height or altitude may fall outside the shape.

Exercise 6D

1 Calculate the area of each parallelogram.

a)

b)

c)

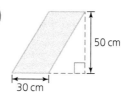

d)

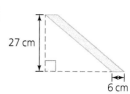

2 Calculate the area of each parallelogram. Remember to use the correct square units.

a)

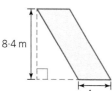

b)

c)

d)

3   A business card is in the shape of a parallelogram with dimensions shown.

Calculate the area of the card.

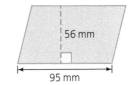

56 mm

95 mm

4   This pattern of identical parallelograms is printed on hazard tape.

Calculate the area of one of the parallelograms in the pattern.

13·8 cm

30 cm

5   A chevron pattern is made from red and white parallelograms.

The base of the white parallelogram is twice the length of the base of the red parallelogram.

a)   Calculate the total area of the red parallelograms shown.

b)   Write down the total area of the white parallelograms shown.

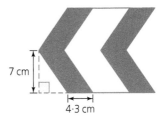

7 cm

4·3 cm

6   A jeweller makes a pair of cufflinks which are in the shape of a parallelogram. Each one has base 14 mm and perpendicular height 23 mm. The visible parallelogram surfaces will receive 3 coats of enamel. Calculate the total area of enamel needed to make the pair of cufflinks.

7   A mirror is in the shape of a parallelogram.

The outside dimensions of the frame are shown. The frame is 15 mm wide all the way around.

a)   Find the length of the base and perpendicular height of the mirror.

b)   Calculate the area of the mirror.

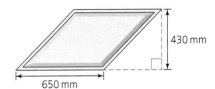

430 mm

650 mm

8   Sketch 3 parallelograms that have a base which is a square number and an area of $72\,\text{cm}^2$.

9   A wind farm covers a parallelogram shaped area as shown.

The owners have permission to increase the area of the windfarm by 20%.

The base length of 16 km and parallelogram shape must be preserved.

Calculate the new perpendicular height of the windfarm.

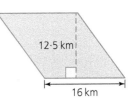

12·5 km

16 km

10   The picture shows a parallelogram shaped puzzle called a tangram.

The small blue square has an area of 1 unit^2.

a)   What is the area of the pink triangle?

b)   What is the area of the green triangle?

c)   What is the area of the whole puzzle?

d)   What fraction of the puzzle is yellow?

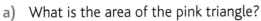

11   Use centimetre-squared paper to draw out a tangram puzzle like the one in question **10**. Take care to match the areas exactly. Cut out the pieces. The rules of the puzzle are simple. Use all the pieces and ensure they are all touching. Try to use all the pieces of your puzzle to make:

a)   an animal   b)   a number   c)   a letter.

Sketch your solutions in your jotter.

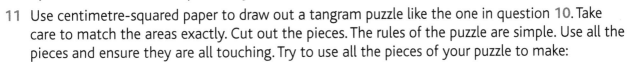

▶ The areas of a rhombus and kite

A **kite** occupies half of the area of the surrounding rectangle.

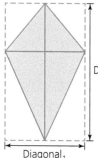

$$\text{Area} = \frac{1}{2} \times \text{diagonal}_1 \times \text{diagonal}_2$$

Diagonal₂

Diagonal₁

This is a kite. A kite has one line of symmetry.

The same formula will also calculate the area of a **rhombus**.

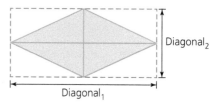

Diagonal₂

Diagonal₁

This is a rhombus. A rhombus has two lines of symmetry.

To find the area of a kite or rhombus:

- Clearly state the formula $A = \frac{1}{2} \times d_1 \times d_2$ or $A = \frac{1}{2}d_1d_2$.
- Substitute the correct values for *diagonal*₁ and *diagonal*₂.
- Calculate and include square units.

Worked examples

1 Calculate the area of the kite.

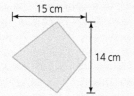
15 cm

14 cm

$$A = \frac{1}{2} \times d_1 \times d_2$$
$$= \frac{1}{2} \times 15 \times 14$$
$$= 105 \, \text{cm}^2$$

We can do the calculation in any order. When one of the numbers is even it is easier to halve that first.

14 × 15 = 210   210 ÷ 2 = 105

or

14 ÷ 2 = 7   15 × 7 = 105

2 Calculate the area of the kite. All lengths are in millimetres.

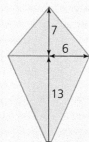
7
6
13

7 + 13 = 20 mm
6 + 6 = 12 mm

$$A = \frac{1}{2} \times d_1 \times d_2$$
$$= \frac{1}{2} \times 20 \times 12$$
$$= 120 \, \text{mm}^2$$

We need to add the shorter dimensions to find the length of the full diagonals.

Remember to include units.

Exercise 6E

1 Calculate the area of each kite or rhombus. Remember to include the correct squared units.

a)

4 m
7 m

b)
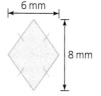
6 mm
8 mm

c)
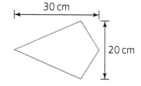
30 cm
20 cm

d)

5 cm
16 cm

e)

 65 mm   30 mm

f) Lengths are in centimetres.

 9   6

g) Lengths are in metres.

 4   7   13

h) Length are in centimetres.

11·4   4·5   6·6

2 Mrs Stirling is deciding which kite to buy.

She wants to buy the kite with the largest area.

Which kite would you recommend Mrs Stirling buys?

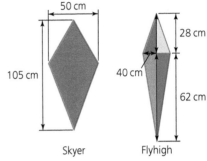

50 cm
105 cm
Skyer

28 cm
40 cm
62 cm
Flyhigh

3 A compass design is formed from 4 identical kites.
The kites are to be coated in luminous paint.
Calculate the total area to be coated in luminous paint.

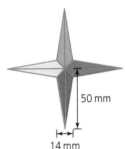

50 mm
14 mm

4 A large neon sign is in the shape of a rhombus.
The internal dimensions are shown. Calculate the area
enclosed by the neon sign.

Give your answer in square metres. Write all lengths in
the same units before you calculate the area.

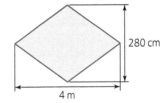

280 cm
4 m

5 In this question make accurate drawings using a ruler.

a) Draw a kite with diagonals 8 cm and 10 cm.

b) Draw 2 more kites and a rhombus which all have the same area as the kite in part **a)**.

6 A kite has one line of symmetry.

When the diagonals do not intersect a V-shaped kite is formed.

a) Calculate the area of the V-shaped kite by subtracting one isosceles
triangle from another.

b) The formula $A = \frac{1}{2} \times d_1 \times d_2$ can also be used for a V-shaped kite.
Use the formula to check your answer to part **a)**.

4 cm
5 cm
10 cm

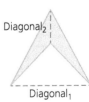

Diagonal$_2$
Diagonal$_1$

7 Sketch 3 V-shaped kites with area = 24 cm².

▶ The area of a trapezium

A trapezium is a quadrilateral with one pair of parallel sides.

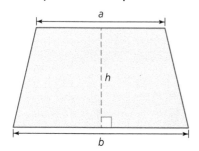

The area of any trapezium can be found using the formula:

$$Area = \frac{1}{2}(a + b)h$$

a and b are the parallel sides and h is the perpendicular height.

To find the area of a trapezium:

- Clearly state the formula: $A = \frac{1}{2}(a+b)h$ or $A = \frac{1}{2} \times (a+b) \times h$.

- Substitute the correct values for the parallel sides a and b and height h.

- Calculate and include square units.

Worked examples

1   Calculate the area of the trapezium.

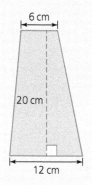

$$A = \frac{1}{2}(a + b)h$$

$$= \frac{1}{2} \times (12 + 6) \times 20$$

$$= \frac{1}{2} \times 18 \times 20$$

$$= 180 \, cm^2$$

Remember to use BODMAS.

Brackets first! $12 + 6 = 18$

Now multiply (any order).

$\frac{1}{2}$ of $20 = 10$ $10 \times 18 = 180$

or

$18 \times 20 = 360$ $360 \div 2 = 180$

2   Calculate the area of the trapezium.

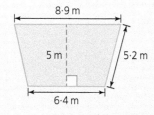

$$A = \frac{1}{2}(a + b)h$$

$$A = \frac{1}{2} \times (8 \cdot 9 + 6 \cdot 4) \times 5$$

$$= \frac{1}{2} \times 15 \cdot 3 \times 5$$

$$= 38 \cdot 25 \, m^2$$

We do not use the sloping length. We need the parallel sides a and b and the height h. The height is the perpendicular distance to the opposite side.

$$\begin{array}{r} 8 \cdot 9 \\ + 6_1 \cdot 4 \\ \hline 15 \cdot 3 \end{array}$$

$$\begin{array}{r} 15 \cdot 3 \\ \times_2 \, _1 5 \\ \hline 76 \cdot 5 \end{array}$$

$$\begin{array}{r} 3 \, 8 \cdot 2 \, 5 \\ 2 \overline{) 7 \, ^1 6 \cdot 5 \, ^1 0} \end{array}$$

Exercise 6F

1 Calculate the area of each trapezium.

a)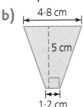
3 m
8 m
7 m

b)
4·8 cm
5 cm
1·2 cm

c)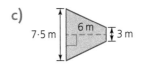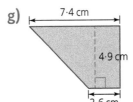
7·5 m
6 m
3 m

d)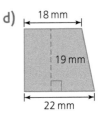
18 mm
19 mm
22 mm

e)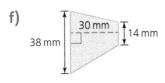
11·5 cm
10 cm
16 cm

f)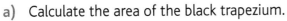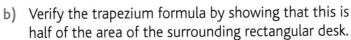
30 mm
14 mm
38 mm

g)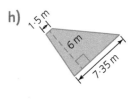
7·4 cm
4·9 cm
2·6 cm

h)
1·5 m
6 m
7·35 m

2 An office desk is inlaid with black marble to form two identical trapezia.

The dimensions are shown.

a) Calculate the area of the black trapezium.

b) Verify the trapezium formula by showing that this is half of the area of the surrounding rectangular desk.

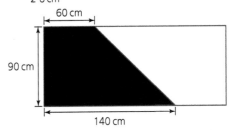
60 cm
90 cm
140 cm

3 A family are considering putting solar panels on their south-facing roof.

The south-facing roof area is in the shape of a trapezium with dimensions shown.

A maximum of 80% of the roof may be covered in solar panels.

Calculate the area available for the solar panels.

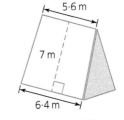
5·6 m
7 m
6·4 m

4 A fishing area is in the shape of a trapezium with the dimensions shown.

Is the area larger than 2500 square kilometres?

108 km
45 km
6 km

5 The lid of a bin designed for recycling is shown.

Calculate the area of the lid in square centimetres.

Write all dimensions in the same units before you calculate the area.

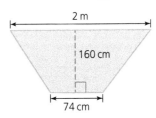
2 m
160 cm
74 cm

6 A helipad on an oil rig is in the shape of a trapezium with parallel sides 14 m and 12 m long.

The area of the helipad is 130 m². Calculate the perpendicular distance between the parallel sides.

▶ Areas of compound shapes

We can join simple shapes together to form a composite or compound shape. To find the area of a compound shape:

● Divide the shape into two or more simple shapes.

● Find the area of each simple shape. Take extra care to identify the correct dimensions.

● Add the simpler areas to find the area of the compound shape.

Worked examples

1 Calculate the area of this shape.

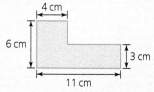

Split the shape into two rectangles.

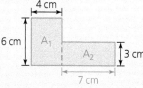

Look carefully at the shapes. We have formed a new rectangle and its length is not 11.

$11 - 4 = 7\,cm$

$A_1 = l \times b$
$= 6 \times 4$
$= 24\,cm^2$

$A_2 = l \times b$
$= 7 \times 3$
$= 21\,cm^2$

Add the simpler areas to find the total composite area.

$A = A_1 + A_2$
$= 24 + 21$
$= 45\,cm^2$

2 A plan of the gable end of a house is shown. Calculate the area of the gable end.

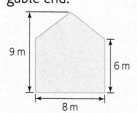

Split this shape into a rectangle and a triangle.

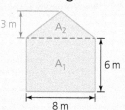

Look carefully. The height of the triangle is 3 m, $9 - 6 = 3\,m$

$A_1 = l \times b$
$= 8 \times 6$
$= 48\,m^2$

$A_2 = \frac{1}{2} \times b \times h$
$= \frac{1}{2} \times 8 \times 3$
$= 12\,m^2$

$A = A_1 + A_2$
$= 48 + 12$
$= 60\,m^2$

There is often more than one way to split the shape and obtain the same answer.

Exercise 6G

1 Calculate the area of each shape. Some have been split for you and a dimension you will need is marked in blue.

a)

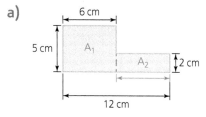

b)

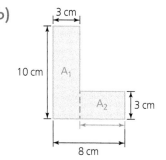

c)

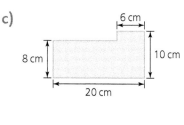

d)

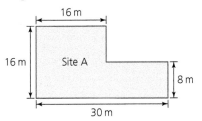

e)

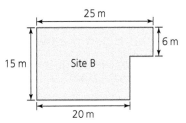

f)

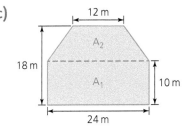

2   Mrs Bryce is looking for land to build a car park. The two plots shown are beside each other and are both the same price. Mrs Bryce will make more money from a larger site. Which site has the larger area?

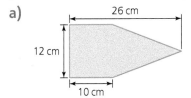

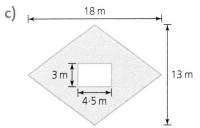

3   Did you notice that **1f)** above is also a trapezium? In these questions think about all the formulae you have learned in this chapter. Calculate the area of each shape.

a)

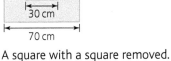

b)

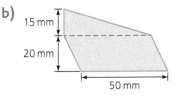

c)

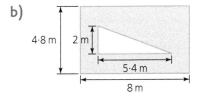

4   In these examples it is easier to find the composite area by subtraction.

Calculate the areas of the outside and inside shapes.

Subtract to find the area which is shaded.

a)

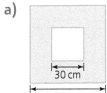

A square with a square removed.

b)

c)

5   An airport plans to create a large arrow on the floor of the departure lounge.

The dimensions are shown.

a)   Calculate the total area of the arrow. Give your answer in square metres.

b)   The cost of installing the floor is £280 per 10 m² or part thereof. Calculate the total cost of the arrow.

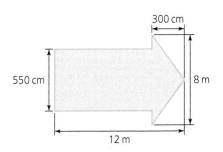

▶ Perimeter and area

Perimeter is the distance around the outside of a shape. To calculate perimeter:

- Find the sum of the lengths around the outside of the shape.
- Use the correct units for length.

Worked example

The diagram shows a symmetrical trapezium.

a) Calculate the perimeter of the shape.

b) Calculate the area of the shape.

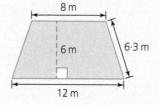

a)

$$\begin{array}{r} 6\cdot3 \\ \times\ 2 \\ \hline 12\cdot6 \end{array}$$

$$\begin{array}{r} 12\cdot6 \\ 12\cdot0 \\ +\ {}_18\cdot0 \\ \hline 32\cdot6 \end{array}$$

There are two sides with length 6·3 m.

When calculating perimeter we **do** use the sloping length.

Perimeter is the length around the shape. We use units of length.

Perimeter = 32·6 m

b)

$$A = \frac{1}{2}(a + b)h$$

$$= \frac{1}{2} \times (8 + 12) \times 6$$

$$= \frac{1}{2} \times 20 \times 6$$

$$= 60\,\text{m}^2$$

When calculating area we **do not** use the sloping length.

Area is the space occupied by the shape. We use square units.

Area = 60 m²

Exercise 6H

1 Calculate the perimeter of each shape.

a)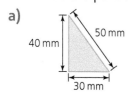

b)

c)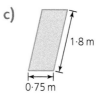

d)
a regular pentagon

e)
curved (arc) length
= 23·55 m

f)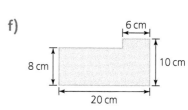

2 The diagrams show the carparks from question **2** in Exercise 6G.

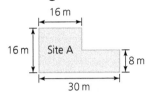

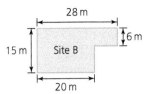

a) Calculate the perimeter of each car park.

Mrs Bryce plans to surround the land she buys with a perimeter fence, leaving a 4 m gap for entry.

The fence costs £40 per metre to construct.

b) Calculate the cost of surrounding each site with fencing.

c) What is the difference in cost between the sites?

3 The diagram shows the dimensions of a triangular patio.

a) Calculate the perimeter of the patio.

b) Calculate the area of the patio.

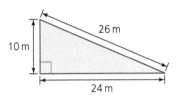

4 The diagram shows a rhombus design from the floor of an Italian villa.

a) Calculate the area of the rhombus.

b) Calculate the perimeter of the rhombus.

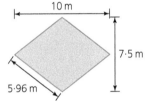

5 The diagram shows the dimensions of a sign.

a) Calculate the perimeter of the sign.

b) Calculate the area of the sign.

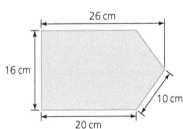

6 The perimeter of this shape is 190 cm.

a) Calculate the length of the side marked x cm.

b) Calculate the area of the shape.

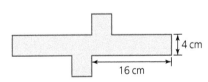

7 This shape consists of two identical L shapes.
The area of the whole shape is 146 cm².
Calculate the perimeter of the shape.

8 **a)** The area of a rectangle is 48 m². The perimeter of the rectangle is 38 m. Find the dimensions of the rectangle.

b) The area of a rectangle is 324 cm². The perimeter of the rectangle is 78 cm. Find the dimensions of the rectangle.

9 How many shapes can you find that have the same numerical value for area and perimeter?

▶ Compound volume

The **volume** of a shape is the amount of 3D space it occupies.

Volume = length × breadth × height
$$V = l \times b \times h$$

Volume is measured in cubic units.

To find the volume of a compound cuboid:

- Split the shape into simpler cuboids.
- Clearly state the formula $V = l \times b \times h$ or $V = lbh$. Take care to substitute the correct dimensions.
- Calculate the compound volume by adding the simpler volumes together. Use cubic units.

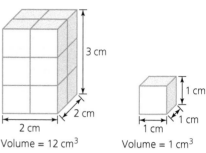

Volume = 12 cm³   Volume = 1 cm³

| Worked example | Calculate the volume of this shape. |

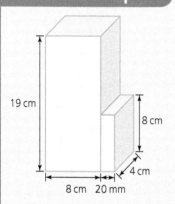

$V_1 = l \times b \times h$
$\quad = 8 \times 4 \times 19$
$\quad = 608\,cm^3$

$$\begin{array}{cc} 19 & 76 \\ \times_3 4 & \times_4 8 \\ \hline 76 & 608 \end{array}$$

We can multiply in any order.

We can see from the diagram that the two cuboids have the same breadth.

20 mm
20 ÷ 10
= 2 cm

$V_2 = l \times b \times h$
$\quad = 2 \times 4 \times 8$
$\quad = 64\,cm^3$

$V = V_1 + V_2$
$\quad = 608 + 64$
$\quad = 672\,cm^3$

The final step is to add the simpler volumes. Remember the cubic units.

Exercise 6I

1   Calculate the volume of each cuboid or compound cuboid.

a)
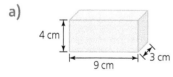

b)
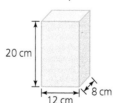

c)
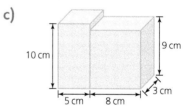

d)
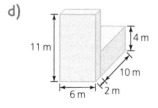

e)
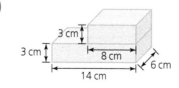

f)
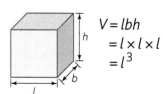

2   Notice the formula $V = l^3$ can be used to find the volume of a cube.
Use the formula $V = l^3$ to calculate the volume of a cube with length:

a)   2 cm   b)   5 cm   c)   9 m   d)   20 mm.

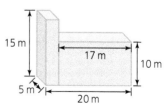

$V = lbh$
$\quad = l \times l \times l$
$\quad = l^3$

➜

3 Calculate the volume of these boxes. Give your answer in the units shown.

Remember to write all dimensions in the correct units before you calculate the volume.

a) cubic millimetres   **b)** cubic centimetres   **c)** cubic metres

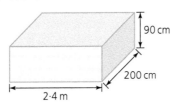

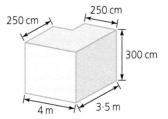

Worked example

The volume of this cuboid is 140 cm³. Find the length.

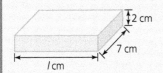

$V = lbh$ Set down the formula.

$140 = l \times 7 \times 2$ Substitute all known dimensions.

$140 = l \times 14$

$b = 10\,cm$

---

4 Calculate the unknown dimension.

a)

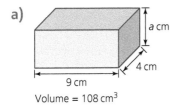

9 cm

Volume = 108 cm³

b)

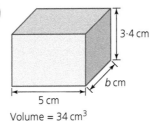

5 cm

Volume = 34 cm³

c)

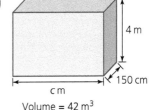

_c_ m

Volume = 42 m³

5 The diagram shows the dimensions of a factory that has three workshops. Workshops A and B have the same length and breadth.

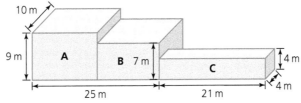

a) Calculate the difference in volume between workshop A and workshop B.

b) Calculate the total volume of all three workshops.

6 A product designer is asked to create a cuboid gift box that has volume 216 cm³.

a) How many different designs for a box with volume = 216 cm³ can you find? Sketch out your designs and label the dimensions.

b) The manufacturer wants to minimise the amount of cardboard used to create the box. Which design would you recommend?

▶ The volume of a prism

A **prism** is a 3D shape with a constant **cross-section**. We can imagine cutting the shape into identical slices.

The space a prism occupies depends on the area of its cross-section (slice) and the height of the prism. The formula is:

Volume = area of cross-section × height

$$V = A \times h$$

To find the volume of a prism:

- Clearly state the formula: $V = A \times h$ or $V = Ah$.
- Substitute the correct values for the cross-sectional area A and height h.
- Calculate and include cubic units.

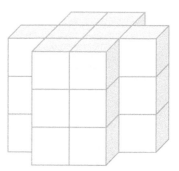

Area (cross-section) = 8 cm²

$$V = A \times h$$
$$= 8 \times 3$$
$$= 24 \, cm^3$$

Worked example Calculate the volume of this prism.

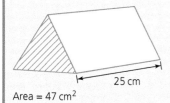

Area = 47 cm²

$$V = A \times h$$
$$= 47 \times 25$$
$$= 1175 \, cm^3$$

```
      4 7
    × 2₃5
    ------
    2 3 5
  + 9 4 0
    ------
  11 7 5
```

This is a triangular prism. The cross-sectional shape is a triangle.

Notice, the 'height' is 25 cm.

The height of a prism is the length between opposite identical faces.

Exercise 6J

1 Calculate the volume of each prism.

a)

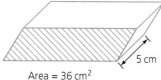

Area = 36 cm²

5 cm

b)

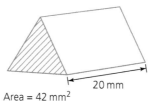

Area = 42 mm²

20 mm

c)

Area = 84 cm²

17 cm

d)

Area = 7·05 m²

3 m

e)

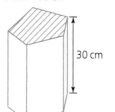

Area = 93 cm²

30 cm

f)

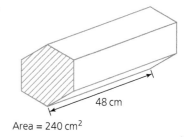

48 cm

Area = 240 cm²

2 Each shape has a constant cross-section. For each one, calculate:

a) the area of the cross-section

b) the volume of the prism.

Shape 1

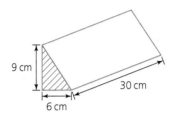

Shape 2

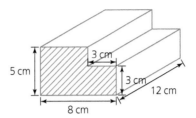

Shape 3

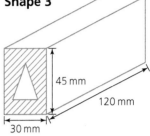

The triangular hole has:
base = 18 mm, height = 25 mm

3 Calculate the unknown values.

a) Volume = 225 cm³

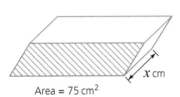

Area = 75 cm²

b) Volume = 240 mm³

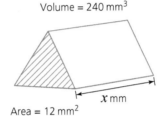

Area = 12 mm²

c) Volume = 450 cm³

Area = x cm²

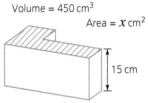

4 The diagram shows a design for a garden water feature.

a) Calculate the total area of the base of the water feature.

b) Calculate the volume of the water feature using the prism formula.

c) Calculate the liquid capacity of the tank in litres.

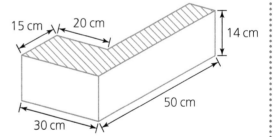

5 A gold bar is in the shape of a trapezium-based prism with the dimensions shown.

a) Calculate the area of the cross-section.

b) Calculate the volume of the gold bar.

c) The bar is fine gold which has a density of 19·3 g per cm³. Calculate the weight of the gold bar.

d) The spot price of gold is given in a traditional measuring system called troy ounces. The spot price one day was £1300 per troy ounce. This is equivalent to approximately £42 per g.

Calculate the value of the gold bar that day.

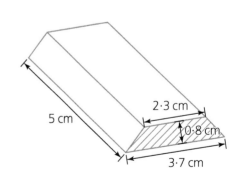

Check-up 👍

1 Copy and complete these tables of formulae.

2D Shape	Area
Rectangle	$A = l \times b$
Square	$A = l^2$
Triangle	$A =$
Rhombus or kite	$A =$
Parallelogram	$A =$
Trapezium	$A =$

3D Shape	Volume
Cuboid	$V =$
Cube	$V = l^3$
Prism	$V =$

2 Calculate the area of each shape.

Put the shapes in order of area from smallest to largest to form a mathematical word.

R

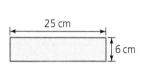
25 cm
6 cm

E

13 cm

A

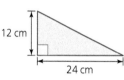
12 cm
24 cm

T

7 cm
15·5 cm

E

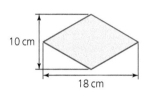
10 cm
18 cm

H

9 cm
15 cm

C

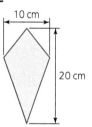
10 cm
20 cm

3 Two roof supports are shown. Which roof support spans the largest area? Remember units must be consistent.

Support 1

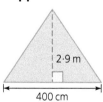
2·9 m
400 cm

Support 2

3 m
350 cm

4 Three designs for the base of a box of chocolates are shown. The manufacturer wants a box base with an area no greater than 900 cm². Which designs are acceptable?

Design 1

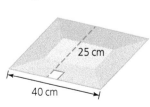
25 cm
40 cm

Design 2

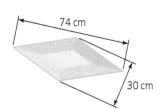
74 cm
30 cm

Design 3

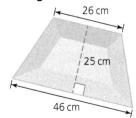
26 cm
25 cm
46 cm

5 A kite is designed in two colours as shown.

Calculate the area of:

a) the whole kite

b) the pink kite

c) the section which is shaded blue.

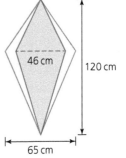

46 cm

120 cm

65 cm

6 An office building has 25 identical rectangular windows with the internal dimensions shown.

The office manager says the window glass covers more than 100m².

Is he correct? Clearly explain your answer.

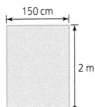

150 cm

2 m

7 An estate agent sketches a floorplan of a studio flat.

The flat has two rectangular areas.

a) Calculate the total area of the flat.

b) Calculate the perimeter of the flat.

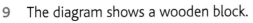

8·4 m

5 m

18 m

4·5 m

8 A recycling bin has three separate compartments.

Each compartment has the same length and breadth.

a) The total length of the bin is 120 cm. Calculate the length of one of the compartments.

b) Calculate the volume of the compost compartment.

c) How much bigger is the compartment for paper?

d) What is the total capacity of the recycling bin? Give your answer in litres.

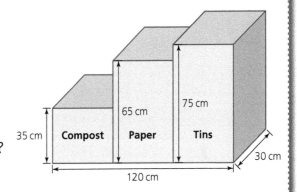

35 cm   Compost   Paper   Tins

65 cm

75 cm

30 cm

120 cm

9 The diagram shows a wooden block.

a) What is the name of this shape?

b) Calculate the area of the cross-section.

c) Calculate the volume of the shape.

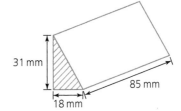

31 mm

85 mm

18 mm

10 A square picture frame is built using four congruent trapezia.

a) Calculate the area of the frame by subtracting the area of one square from another.

b) Calculate the area of one of the trapezia.

c) Find the width of the frame marked *x* mm.

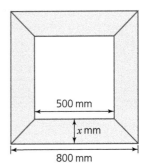

500 mm

x mm

800 mm

7 Expressions and equations

▶ Gathering like terms

We use letters in mathematics to represent an unknown quantity. The letter is called a variable.

- There is no need to write the number 1 in front of a variable: $1x = x$, $1p = p$, $-1k = -k$.
- Do not write the multiplication sign between a number and a variable: $2 \times x = 2x$, $3 \times y = 3y$.
- When a variable appears more than once in an expression simplify to a single term.

| Worked examples | Simplify these expressions: |

1 $x + x + x$

$= 3x$ (3 lots of x)

2 $m + m + m - m$

$= 2m$ (2 lots of m)

3 $z + 5z$

$= 6z$ (6 lots of z)

4 $3p + 5p - 2p$

$= 6p$

5 $3t + 6t - 8t$

$= t$

6 $w + 2w - 8w$

$= -5w$

Exercise 7A

1 Simplify:
   a) $x + x + x + x$ b) $y + y + y$ c) $m + m + m + m - m$ d) $k + k + k + k - k - k - k$

2 Write as a single term:
   a) $x + 3x$ b) $x + 5x$ c) $y + 6y$ d) $z + 10z$ e) $8m + m$ f) $4w + 3w$
   g) $2v + v$ h) $6h + 9h$ i) $20g + 30g$ j) $x + 4x + 2x$ k) $3q + q + 2q$ l) $5e + 7e + 3e$

3 Write as a single term:
   a) $5x - x$ b) $7y - 2y$ c) $3p - 2p$ d) $12m - 10m$
   e) $20f - 14f$ f) $16t - 10t$ g) $7y + 2y - 3y$ h) $2d + 13d - 8d$
   i) $14l + l - 2l$ j) $20w - 10w + 5w$ k) $13j - 5j + j$ l) $10y - 3y - 7y$

4 Write each expression as a single term. (Your answer should be negative in each case.)
   a) $x - 3x$ b) $4k - 8k$ c) $10z - 11z$ d) $4t + 3t - 10t$
   e) $6m + 2m - 13m$ f) $14j + 13j - 33j$ g) $5t - 2t - 6t$ h) $10p - 8p - 7p$
   i) $15x - 10x - 18x$ j) $6h - 5h - 3h$ k) $8y - 10y + y$ l) $17x - 25x + 4x$

Consider the expression $2x + 7y + 4x - 3y + 5k + 1$.

As the expression contains a mixture of variables and a single number, we gather like terms to simplify.

→ $2x + 4x$ are like terms because they contain the variable x and simplify to $+ 6x$.

→ $+ 7y - 3y$ are like terms because they contain the variable y and simplify to $+ 4y$.

→ There is only one term involving k and one single number, so they do not have any like terms and remain unchanged.

→ It is important to remember that the sign in front of a term ($+$ or $-$) refers to that term only, not everything that comes after it. Here, the minus only acts on the $3y$ term, not on the $+ 5k$ and $+ 1$ terms after it.

Thus, we can simplify this expression to $2x + 7y + 4x - 3y + 5k + 1$

$$= 6x + 4y + 5k + 1$$

When gathering like terms:

- Identify and simplify like terms.

- Leave any terms without like terms unchanged.

Worked examples Simplify these expressions:

1 $2x + 3x + 6y$

   $2x + 3x + 6y$

   $= 5x + 6y$

   The $2x$ and $+ 3x$ are like terms and simplify to $5x$.
   The $+ 6y$ term has no like terms so it remains unchanged.

2 $4k - 8m + 2k - m - p$

   $4k - 8m + 2k - m - p$

   $= 6k - 9m - p$

   The $4k$ and $+ 2k$ are like terms and the $- 8m$ and the $- m$ are like terms that simplify to $6k$ and $- 9m$, respectively. The $- p$ term has no like terms so it remains unchanged.

3 $3p + 5 - p + 2$

   $3p + 5 - p + 2$

   $= 2p + 7$

4 $6k + 9h + 3 + 2k - 10h$

   $6k + 9h + 3 + 2k - 10h$

   $= 8k - h + 3$

5 $x + 12 - 4w - 3x + w - 5$

   $x + 12 - 4w - 3x + w - 5$

   $= -2x + 7 - 3w$

6 $8 - 2x + 9y + x - 5m - 3y$

   $8 - 2x + 9y + x - 5m - 3y$

   $= 8 - x + 6y - 5m$

Exercise 7B

1 Gather like terms to simplify:

   a) $5x + 3x + y$
   b) $6y + 3 + y$
   c) $9c - 5c + 1$
   d) $2 + 8h - 3h$

2 Simplify:

   a) $2x + 4x + 3y + 7y$
   b) $9x - 3x + 6y + 2y$
   c) $10m + 3n + m - n$
   d) $5k + 6p - 2k + 3p$
   e) $12z + 3d - 2z - d$
   f) $7x + 6 - 2x + 1$
   g) $12g + 10f + g - 4f$
   h) $6t + 5u - 2t - u$

3 Simplify:

   a) $6x + 2y + 3x + y + 8$
   b) $4m + 7n + 10m - 2n + 4$
   c) $5r + 3s + 7s - 2r - p$
   d) $10h + 14k - 6h - k + p$
   e) $4x - 9x + 3y$
   f) $10p + 7q - 12p + q$
   g) $2r - 3t + r - t$
   h) $5d - 3f - 6d + 2f + 7$
   i) $-x - 7y - 4x + 2y - 6$

4 a) The bubbles contain four expressions. Simplify each one and decide which is the odd one out.

   $2m + 3n - 4m$
   $5n - 6m - 2n + m$
   $3n - 5m - 3m + 6m$
   $4m - 6m + 3n$

   $8m + 4n - 2m$
   $8n + 6m - n - 3n$
   $7n - m - 3n + 7m$
   $5n - 5m - 2n + m$

   $10m - 6n - m$
   $-5n - m - n + 9m$
   $-8n + 6m + 2n + 2m$
   $10m - 2m - 6n$

   b) Find the sum of your answers to part a).

Sometimes the terms in an expression can have powers or include more than one variable.

Worked examples   Simplify these expressions:

1    $4x^2 + 8x^2 + 6x - 2x$
     $4x^2 + 8x^2 + 6x - 2x$
     $= 12x^2 + 4x$

The $4x^2$ and $+ 8x^2$ are like terms and simplify to $12x^2$. The $+ 6x$ and the $- 2x$ are like terms and simplify to $+ 4x$. Note that whilst the x^2 and x contain the same letter they are not like terms!

2    $5ab + 4a + 2b - ab$
     $5ab + 4a + 2b - ab$
     $= 4ab + 4a + 2b$

The $5ab$ and $- ab$ are like terms and simplify to $4ab$. The $+ 4a$ and $+ 2b$ have no like terms as there are no other terms containing the single variables so they remain unchanged.

3    $8pq + 2m - 3qp - 6m$
     $8pq + 2m - 3qp - 6m$
     $= 5pq - 4m$

The $8pq$ and $- 3qp$ are like terms as they both contain the variables p and q. We can rewrite the $- 3qp$ as $- 3pq$ to have the variables in the same order as the $8pq$.

Exercise 7C

1   Gather like terms to simplify:
   a) $x^2 + 4x^2$
   b) $5y^2 + 3y^2$
   c) $9m^2 - 3m^2$
   d) $10p^2 - 5p^2$
   e) $3x^2 + 6x^2 - 2x^2$
   f) $12y^2 - 4y^2 + y^2$

2   Simplify:
   a) $3xy + 5xy$
   b) $2mn + 8mn$
   c) $7pq - 3pq$
   d) $11ab - 9ab$
   e) $15uv + 2uv - 5uv$
   f) $20xy - 6xy - 8xy$

3   Simplify:
   a) $5x^2 + 3y^2 + x^2 + 4y^2$
   b) $7p^2 + 10q^2 + p^2 - 3q^2$
   c) $8u^4 + 5v^3 - 2u^4 - v^3$
   d) $4mn + 3pq + 2mn + pq$
   e) $12ab + 9cd + 3ab - 4cd$
   f) $10pqr + 2rst - 5pqr - rst$

4   Simplify:
   a) $6m^2 - 3m - 2m^2 + m$
   b) $5x^2 - 8x - 3x^2 - 4x$
   c) $-4k^2 + k + 5k^2 - 2k$
   d) $17xy - 4yz - 10xy + 3yz$
   e) $4ab - 9cd + 8ba - dc$
   f) $-abc - 3efg + 4bca + gfe$

5   Match the cards that have the same expression on them. Write the matching card letters (in red) with the capital letter first.

A $7xy + 3x$
   $- 4x + 12$

b $2x + xy + 12$
   $- 3x - 6xy$

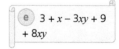
e $3 + x - 3xy + 9$
   $+ 8xy$

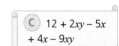
C $12 + 2xy - 5x$
   $+ 4x - 9xy$

n $8xy + x + 10$
   $+ 2 - 15xy$

F $6x - 1 + 13$
   $+ 5xy - 5x$

S $14 - 12xy + x$
   $+ 5xy - 2$

g $3 - x - 2xy + 9$
   $+ 9xy$

a $10 - 7xy + 2$
   $- 3x + 2x$

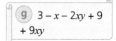
P $7x - 5xy + 10$
   $- 8x + 2$

Your answers should represent elements from the periodic table. Which elements have you found?

Worked example To complete a pyramid, add together the expressions in two adjacent boxes to obtain the expression in the box above.

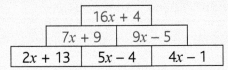

For example, we obtain the expression in the first box in the middle row by adding together the expressions in the two boxes below it and simplifying.

$2x + 13 + 5x − 4 = 7x + 9$

6 Copy and complete:

a)
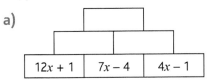

b)
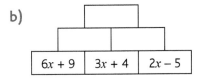

c)
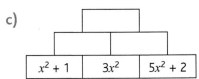

d)
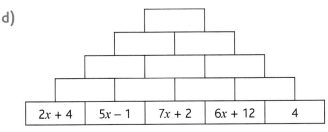

7 Copy and complete:

a)
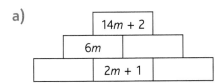

b)
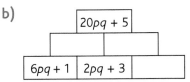

c)
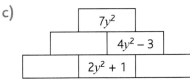

d)
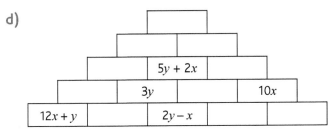

8 Make some of your own pyramids and test a classmate with them.

9 Write a single expression (e.g. $2xy$, 5, $7ab$, ...) in each box to make the statements true. You cannot change the sign in front of a box.

a) $2x + \boxed{} + 6x − 1 = 3x + 5 + \boxed{} − 2$

b) $6x^2 + 4x^2 + 13 + \boxed{} = \boxed{} + 3x^2 + 16$

c) $12ab + 9 − \boxed{} + 2 = 15 − \boxed{} − ab$

d) $p^2 + 3q^2 + 8 − \boxed{} = 13 + 2p^2 − \boxed{} − q^2 − \boxed{}$

e) $pqr + 5pq − 2qr − \boxed{} − q + 2 = 5 − \boxed{} − \boxed{} + 3pqr + qr + 2pq − \boxed{} − \boxed{} + 4q$

▶ Evaluating expressions

We can **evaluate** (find a value for) an expression when the variables are assigned numeric values.

- **Substitute** the number in for the corresponding variable.
- Use **BODMAS** to evaluate the expression.

Worked examples

1 If $x = 3$ and $y = 2$, evaluate:

  a) $x + 6$

     $= 3 + 6 \rightarrow$ sub in $x = 3$

     $= 9 \quad \rightarrow$ simplify

  b) $y - 3$

     $= 2 - 3 \rightarrow$ sub in $y = 2$

     $= -1 \quad \rightarrow$ simplify

  c) $x + y$

     $= 3 + 2 \rightarrow$ sub in x and y

     $= 5 \quad \rightarrow$ simplify

2 Given $p = 4$, $q = 5$ and $r = 10$, evaluate:

  a) $2p$

     $= 2 \times 4$

     $= 8$

  b) $-3q$

     $= -3 \times 5$

     $= -15$

  c) $1 - 3p$

     $= 1 - 3 \times 4$

     $= 1 - 12$

     $= -11$

  d) $2p + 3q - pr$

     $= 2 \times 4 + 3 \times 5 - 4 \times 10$

     $= 8 + 15 - 40$

     $= -17$

Exercise 7D

1 Substitute $x = 2$ and $y = 8$ into each expression and evaluate it.

  a) $x + 3$
  b) $x - 6$
  c) $y + 5$
  d) $y - 10$

  e) $x + y$
  f) $y - x$
  g) $y + y + y - x$
  h) $6 - y + x$

2 Given $m = 3$ and $n = 6$, evaluate each expression.

  a) $4m$
  b) $5m$
  c) $2n$
  d) $7n$

  e) $-2m$
  f) $-3n$
  g) mn
  h) $10nm$

3 For $g = 4$ and $h = 5$, evaluate each expression.

  a) $2g + 1$
  b) $3g - 5$
  c) $5g + 6$
  d) $2g - 12$

  e) $3g - 20$
  f) $4h + 3$
  g) $6h - 1$
  h) $2h - 15$

  i) $3h - 22$
  j) $3 + 2g$
  k) $4 + 5g$
  l) $10 - 3g$

4 Given $a = 2$, $b = 5$ and $c = 6$, evaluate each expression.

  a) $2a + b$
  b) $3a - c$
  c) $2b + 3a$
  d) $10a - 3c$
  e) $5a + 4c$
  f) $3a + 2b - 4$

  g) $1 + 3c - 2b$
  h) $12 + 3a + 2c$
  i) $5b - 7a + 5$
  j) $2 + 3a - c$
  k) $10 + ab - 2c$
  l) $cb - ac$

5 In this question, $x = 2$, $y = 3$ and $z = 4$.

Convert the answer to each expression you are given to the corresponding letter in the alphabet. Then rearrange your letters to make a word. The table will help you.

A	B	C	D	E	F	G	H	I	J	K	L	M	N	O	P	Q	R	S	T	U	V	W	X	Y	Z
1	2	3	4	5	6	7	8	9	10	11	12	13	14	15	16	17	18	19	20	21	22	23	24	25	26

→

Here is an example completed for you to make a capital city:

$4z + x - y$
$= 4 \times 4 + 2 - 3$
$= 15$ which is the
letter O

$y + x$
$= 3 + 2$
$= 5$ which is the
letter E

$3z + 3x$
$= 3 \times 4 + 3 \times 2$
$= 18$ which is the
letter R

$5x + y$
$= 5 \times 2 + 3$
$= 13$ which is the
letter M

Rearranging the letters O, E, R and M gives the capital city ROME.

a) Find the name of two capital cities from these expressions.

$2z + 1$	$yz - 5x$	$6y - z$	$10x - 1$	$4z + x$	$x + y + z$
$4z - 2x$	$2x - y$	$yz + 2x$	$3y - 2x$	$4y + xy$	

b) Write expressions of your own which will give a different capital city. Test a classmate with them.

Worked example   1 For $x = 3$ and $y = 4$, evaluate

a) x^2
$= 3^2$
$= 3 \times 3$
$= 9$

b) $2y^2$
$= 2 \times y^2$
$= 2 \times y \times y$
$= 2 \times 4 \times 4$
$= 32$

c) $(2y)^2$
$= (2 \times 4)^2$
$= (8)^2$
$= 8 \times 8$
$= 64$

Be very careful here!

In b) you have 2 lots of y^2. It is only the y that is squared.

In c) the whole of $(2y)$ is squared.

Note that the answers are different with and without the brackets.

d) $\dfrac{6y - x}{x}$

$= \dfrac{6 \times 4 - 3}{3}$

$= \dfrac{24 - 3}{3}$

$= \dfrac{21}{3}$

$= 7$

---

6 Substitute and evaluate each expression when $m = 2$ and $n = 5$.
   a) m^2
   b) n^2
   c) m^3
   d) n^4
   e) m^5
   f) $3m^2$
   g) $(3m)^2$
   h) $2n^2$
   i) $(2n)^2$
   j) $4m^3$
   k) $(4m)^3$
   l) $6n^3$

7 Substitute $x = 2$ and $y = 4$ into each fraction and evaluate it.
   a) $\dfrac{x + y}{3}$
   b) $\dfrac{3y + x}{2}$
   c) $\dfrac{5y + 2x}{x}$
   d) $\dfrac{1 + 4x}{y - 1}$

8 Evaluate each expression, given that $m = 5$ and $n = 10$.
   a) $3 + 2m^2$
   b) $100 - (n^2 - m)$
   c) $3n^2 - 10m$
   d) $16 + \frac{1}{2}$ of n^3

9 Using the same method as question 5, when $x = 2$, $y = 3$ and $z = 4$ and each answer corresponds to a letter in the alphabet, find the names of two planets in the Solar system.

$5z - 6$	$4x - y$	$2z - (x + y)$	$10x + z - y$	$y^2 + 2z + 1$
	$xyz + 1$	$z^2 + y + 2$	$\dfrac{(xy)^2}{2}$	$\frac{4}{5}$ of $(x + y)^2$
$4x + 2z - y$	x^2z	$\frac{1}{5}$ of $(z^2 + y^2)$	$3x^2 - 3y + x$	$\dfrac{4y^2}{x} - x^2$

Take care when substituting negative numbers. It can be helpful to write brackets around them before simplifying.

Worked examples   1 Given $p = -2$, $q = 3$ and $r = -7$, evaluate

a) $p + q$
   $= (-2) + 3$
   $= 1$

b) $q + r$
   $= 3 + (-7)$
   $= 3 - 7$
   $= -4$

c) $r - 8p$
   $= (-7) - 8 \times (-2)$
   $= (-7) - (-16)$
   $= -7 + 16$
   $= 9$

d) $5pq$
   $= 5 \times (-2) \times 3$
   $= -30$

e) $12 - 4p + 3r$
   $= 12 - 4 \times (-2) + 3 \times (-7)$
   $= 12 - (-8) + (-21)$
   $= 12 + 8 - 21$
   $= -1$

f) $\dfrac{2p - q}{r}$

   $= \dfrac{2 \times (-2) - 3}{-7}$

   $= \dfrac{-7}{-7}$

   $= 1$

Exercise 7E

1 Substitute $x = -1$, $y = 2$ and $z = -3$ into each expression and evaluate it.

   a) $x + 6$ b) $y + z$ c) $x + z$ d) $x - y$ e) $y - z$ f) $x - z$

   g) $z - x + y$ h) $3x$ i) $2z$ j) $-5x$ k) xy l) xyz

2 Evaluate each expression, given that $m = -5$ and $n = -8$.

   a) $2m + 1$ b) $4n + 2$ c) $6m - 3$ d) $2n - 4$ e) $5n - 3m$ f) $4m + 3n$

3 Mr Donaldson has a game to decide which of his four groups of pupils leaves the classroom first. Each group picks one of these cards before he tells them values for m and n. The group with the highest number is first to leave and the group with the lowest number get out last.

   | A | $3m + 4n$ | | B | $5m + 6n$ | | C | $2m + 3n$ | | D | $4m + 5n$ |

   Which card will let the group go out:

   a) first if Mr Donaldson chooses $m = -4$ and $n = 2$ b) last when he chooses $m = -5$ and $n = 3$?

4 Evaluate each expression when $a = -1$, $b = 4$ and $c = -6$.

   a) $\dfrac{a + 5}{2}$ b) $\dfrac{5c}{3}$ c) $\dfrac{b + c}{a}$ d) $\dfrac{3a + 2c}{b + 1}$

5 When the correct set of values in column A is substituted into an expression in column B, it gives a value in column C. Find the correct combinations for this to be true.

Column A	Column B	Column C
$m = 2$, $n = -3$, $p = 6$	$-8 + 3pm + 2m$	-12
$m = -4$, $n = 10$, $p = -2$	$\frac{1}{2}$ of $mp + 6n$	-56
$m = 2$, $n = -1$, $p = 4$	$7mnp$	8

It is also important to use brackets when evaluating powers of negative numbers.
Note that $-3^2 \neq (-3)^2$.

-3^2

$= -3 \times 3$ Only the number 3 is squared,

$= -9$ the negative is not squared.

$(-3)^2$

$= (-3) \times (-3)$ Negative 3 is squared due to the bracket.

$= 9$

Worked examples   1 Evaluate each expression, given that $m = -4$ and $n = -5$.

a) m^2

$= (-4)^2$

$= (-4) \times (-4)$

$= 16$

b) $2m^2$

$= 2 \times (-4)^2$

$= 2 \times (-4) \times (-4)$

$= 32$

c) $(2m)^2$

$= (2 \times (-4))^2$

$= (-8)^2$

$= (-8) \times (-8)$

$= 64$

d) mn^3

$= (-4) \times (-5)^3$

$= (-4) \times (-5) \times (-5) \times (-5)$

$= 500$

---

6   Substitute $p = -2$ and $q = -3$ into these expressions to evaluate them.

a) p^2 b) $3p^2$ c) $(3p)^2$ d) q^2 e) $5q^2$ f) $(5q)^2$

g) p^2q h) q^2p i) $2q^2 + 5p^2$ j) p^3 k) $4p^3$ l) $(4p)^3$

7   Copy this number crossword into your jotter. Use $a = -5$, $b = -6$ and $c = 10$ to evaluate the clues and complete the crossword.

Across

1   $ab + c^2$

3   $(2b)^2 + a$

5   $10c^2 + 84 + 4a$

8   $c^3 + 7b^2$

10   $-3b^3$

12   $a^2c + 50$

Down

2   $abc + 11$

3   $4b^2$

4   $(ab)^2$

6   $7c^2 + b - 2$

7   $b - 3ac$

8   $-a^2b$

9   $\dfrac{c^3 - 11b}{2}$

10   $-10b$

11   $c^2 - \dfrac{1}{2}$ of ab

8   For each expression find two values of x that can be substituted to give the required answer.

	In	Expression	Out
a)	$x = ?$	$8x^2 - 3$	A prime number between 25 and 30
b)	$x = ?$	$(4x)^2 - 4$	The lowest common multiple of 4, 5 and 7
c)	$x = ?$	$68 - 2x^2$	The highest common factor of 54 and 72
d)	$x = ?$	$\dfrac{x^3}{100}$	The x value you put in

▶ Balancing equations

What will happen if you take a sphere from one side of this balance?

You should see that all the spheres will fall off.

However, if you take one sphere from each side, it will stay balanced (equal).

An equation has two sides that are equal. When solving an equation, we must make sure both sides always stay equal. This will only happen if every time you make a change to one side you do exactly the same to the other side. This technique is called balancing.

- Perform balancing steps until you have a value for the variable (see worked examples).
- Check your solution by substituting the value back into the original equation.

Worked examples   Solve the equations:

1. $x + 3 = 7$
   $\quad -3 \quad -3$

   $\quad\quad x = 4$

   Check:
   LHS = 4 + 3 = 7
   RHS = 7 ✓

   To balance the + 3, subtract 3 from both sides.

   As the left-hand side (LHS) equals the right-hand side (RHS), $x = 4$ is correct.

2. $x + 8 = 10$
   $\quad -8 \quad -8$

   $\quad\quad x = 2$

   Check:
   LHS = 2 + 8 = 10
   RHS = 10 ✓

   To balance the + 8, subtract 8 from both sides.

3. $x - 1 = 5$
   $\quad +1 \quad +1$

   $\quad\quad x = 6$

   Check:
   LHS = 6 − 1 = 5
   RHS = 5 ✓

   To balance the − 1, add 1 to both sides.

4. $x - 12 = 4$
   $\quad + 12 \quad + 12$

   $\quad\quad x = 16$

   Check:
   LHS = 16 − 12 = 4
   RHS = 4 ✓

   To balance the − 12, add 12 to both sides.

5. $x + 4 = 2$
   $\quad -4 \quad -4$

   $\quad\quad x = -2$

   Check:
   LHS = − 2 + 4 = 2
   RHS = 2 ✓

   To balance the + 4, subtract 4 from both sides. (The answer can be negative.)

6. $x - 5 = -11$
   $\quad +5 \quad +5$

   $\quad\quad x = -6$

   Check:
   LHS = − 6 − 5 = −11
   RHS = −11 ✓

   To balance the − 5, add 5 to both sides.

In the remaining worked examples we will not show the check stage however you should continue to check your answers, either written or mentally, to ensure you have obtained the correct answer.

Exercise 7F

1   Solve these equations by subtracting a number from both sides.
   a) $x + 2 = 6$ b) $x + 5 = 12$ c) $x + 1 = 4$ d) $x + 3 = 11$
   e) $x + 9 = 14$ f) $x + 4 = 20$ g) $x + 8 = 10$ h) $x + 6 = 25$

2   Solve these equations by adding a number to both sides.
   a) $x - 5 = 1$ b) $x - 2 = 10$ c) $x - 3 = 8$ d) $x - 1 = 13$
   e) $x - 7 = 6$ f) $x - 4 = 11$ g) $x - 6 = 7$ h) $x - 8 = 0$

3   Solve these equations:
   a) $x + 3 = 9$ b) $x + 2 = 10$ c) $x + 7 = 8$ d) $x - 9 = 6$
   e) $x - 4 = 12$ f) $x - 6 = 3$ g) $x - 1 = 18$ h) $x + 1 = 14$
   i) $x - 8 = 4$ j) $x - 10 = 5$ k) $x + 3 = 12$ l) $x - 23 = 1$
   m) $x + 11 = 47$ n) $x - 19 = 11$ o) $x - 25 = 15$ p) $x + 10 = 10$

4   Solve (taking care with negative numbers):
   a) $x + 5 = 1$ b) $x + 10 = 3$ c) $x + 12 = 4$ d) $x + 8 = 5$
   e) $x + 9 = -2$ f) $x + 13 = -3$ g) $x + 20 = -4$ h) $x + 30 = -1$
   i) $x - 8 = -5$ j) $x - 10 = -2$ k) $x - 5 = -14$ l) $x - 4 = -20$
   m) $x - 7 = -12$ n) $x - 14 = -20$ o) $x - 12 = -13$ p) $x - 50 = -90$

Worked examples   Solve these equations:

1   $2x = 16$ Recall that $2x$ means $2 \times x$ 2 $7x = 63$ 3 $5x = -75$
   $\div 2 \quad \div 2$ so to balance the 2, divide $\div 7 \quad \div 7$ $\div 5 \quad \div 5$
   $x = 8$ both sides by 2. $x = 9$ $x = -15$

5   Solve these equations:
   a) $3x = 12$ b) $6x = 42$ c) $9x = 27$ d) $5x = 45$
   e) $4x = 48$ f) $10x = 150$ g) $8x = 112$ h) $2x = 134$

6   Solve these equations (taking care with negative numbers):
   a) $6x = -18$ b) $5x = -55$ c) $8x = -64$ d) $10x = -40$
   e) $7x = -63$ f) $12x = -144$ g) $3x = -150$ h) $4x = -116$

7   Solve these equations (using the two-step division method from page 9):
   a) $20x = 100$ b) $40x = 320$ c) $30x = 210$ d) $90x = 360$
   e) $80x = 880$ f) $70x = 630$ g) $50x = 650$ h) $60x = 1080$

When there is a combination of addition/subtraction and multiplication balance by adding or subtracting first.

Always finish a solution with the variable on the left-hand side.

Worked examples Solve these equations:

1 $2x + 3 = 15$
   $\quad -3 \quad -3$
   $\quad 2x = 12$
   $\quad \div 2 \quad \div 2$
   $\quad x = 6$

2 $6x - 1 = 29$
   $\quad +1 \quad +1$
   $\quad 6x = 30$
   $\quad \div 6 \quad \div 6$
   $\quad x = 5$

3 $8x - 2 = -74$
   $\quad +2 \quad +2$
   $\quad 8x = -72$
   $\quad \div 8 \quad \div 8$
   $\quad x = -9$

4 $25 = 2x + 9$
   $\quad -9 \quad \quad -9$
   $\quad 16 = 2x$
   $\quad \div 2 \quad \div 2$
   $\quad 8 = x$
   $\quad x = 8$

Exercise 7G

1 Solve these equations.

  a) $3x + 1 = 7$
  b) $5x + 2 = 22$
  c) $7x + 5 = 26$
  d) $4x + 9 = 29$
  e) $6x + 1 = 43$
  f) $8x + 1 = 9$
  g) $10x + 2 = 32$
  h) $7x + 18 = 39$
  i) $2x + 67 = 89$
  j) $3x + 15 = 90$
  k) $89 = 5x + 24$
  l) $92 = 4x + 12$

2 Solve these equations.

  a) $3x - 5 = 22$
  b) $5x - 4 = 31$
  c) $8x - 3 = 13$
  d) $7x - 1 = 48$
  e) $6x - 10 = 44$
  f) $4x - 9 = 11$
  g) $2x - 18 = 42$
  h) $3x - 7 = 50$
  i) $7x - 23 = 54$
  j) $6x - 19 = 41$
  k) $85 = 5x - 20$
  l) $82 = 8x - 14$

3 Solve these equations (be careful with negatives).

  a) $2x + 8 = 2$
  b) $5x + 17 = 7$
  c) $6x + 26 = 2$
  d) $4x + 31 = 3$
  e) $8x + 40 = -8$
  f) $3x + 12 = -9$
  g) $10x - 2 = -12$
  h) $9x - 14 = -32$
  i) $2x - 24 = -50$
  j) $7x - 20 = -41$
  k) $-40 = 4x - 16$
  l) $-19 = 5x - 74$

In the following examples, we balance the algebraic terms first.

Worked examples Solve these equations.

1 $6x = 4x + 18$
   $\quad -4x \quad -4x$
   $\quad 2x = 18$
   $\quad \div 2 \quad \div 2$
   $\quad x = 9$

2 $18y = 140 - 2y$
   $\quad +2y \quad \quad +2y$
   $\quad 20y = 140$
   $\quad \div 20 \quad \div 20$
   $\quad y = 7$

3 $3m - 12 = 6m$
   $\quad -3m \quad \quad -3m$
   $\quad -12 = 3m$
   $\quad \div 3 \quad \div 3$
   $\quad -4 = m$
   $\quad m = -4$

4 Solve these equations.

a) $2x = x + 6$
b) $4y = 2y + 10$
c) $10p = 3p + 56$
d) $12k = 4k + 16$
e) $9y = 5y + 36$
f) $7m = 3m + 20$
g) $14x = 8x + 12$
h) $20p = 11p + 72$
i) $32m = 29m + 72$
j) $40y = 30y + 80$
k) $53r = 13r + 120$
l) $65w = 45w + 680$

5 Solve these equations.

a) $3x = 15 - 2x$
b) $7w = 16 - w$
c) $5k = 45 - 4k$
d) $2m = 42 - 5m$
e) $p = 77 - 10p$
f) $4n = 90 - 6n$
g) $14p = 90 - 16p$
h) $23x = 240 - 17x$
i) $41q = 450 - 9q$
j) $37p = 840 - 33p$
k) $69x = 990 - 21x$
l) $93y = 6000 - 7y$

6 Solve these equations.

a) $8x - 4 = 10x$
b) $7m - 15 = 10m$
c) $3p - 21 = 10p$
d) $5k - 52 = 9k$
e) $r - 200 = 6r$
f) $12q = 6q - 42$
g) $15y = 7y - 96$
h) $32g = 2g - 390$
i) $-47x = 1260 + 13x$
j) $-64w = 3780 + 26w$
k) $-24k = 640 - 4k$
l) $-120m = 900 - 20m$

It is often necessary to balance both constants and variables within an equation.

Worked examples   Solve these equations.

1
$$3x + 5 = 2x + 9$$
$$-2x \quad -2x$$
$$x + 5 = 9$$
$$-5 \quad -5$$
$$x = 4$$

2
$$5y + 8 = 42 + 3y$$
$$-3y \qquad -3y$$
$$2y + 8 = 42$$
$$-8 \ -8$$
$$2y = 34$$
$$\div 2 \ \div 2$$
$$y = 17$$

3
$$6x - 7 = x + 48$$
$$-x \quad -x$$
$$5x - 7 = 48$$
$$+7 \ +7$$
$$5x = 55$$
$$\div 5 \ \div 5$$
$$x = 11$$

4
$$2m + 6 = 8m - 24$$
$$-2m \qquad -2m$$
$$6 = 6m - 24$$
$$+24 \qquad +24$$
$$30 = 6m$$
$$\div 6 \ \div 6$$
$$5 = m$$
$$m = 5$$

Exercise 7H

1 Solve these equations.

a) $4x + 6 = 2x + 10$
b) $3y + 7 = y + 17$
c) $6m + 9 = 2m + 13$
d) $8w + 10 = 3w + 30$
e) $10k + 1 = 7k + 19$
f) $9q + 3 = 2q + 52$
g) $8p + 15 = 3p + 90$
h) $7h + 1 = 4h + 58$
i) $6k + 8 = k + 128$

2 Solve these equations.

a) $9x - 6 = 4x + 9$
b) $5m - 2 = 3m + 6$
c) $7y - 9 = 5y + 1$
d) $11k - 13 = 8k + 5$
e) $9x - 8 = 3x + 46$
f) $10w - 13 = 2w + 83$
g) $13p - 22 = 4p + 113$
h) $18m - 51 = 8m + 29$
i) $16y - 47 = 9y + 100$

3 Solve these equations.

a) $4r + 12 = 18 + 2r$
b) $7p + 9 = 21 + p$
c) $8x - 5 = 15 + 3x$
d) $12f - 14 = 49 + 5f$
e) $6m + 25 = 9m + 4$
f) $2y + 68 = 10y + 20$
g) $3x + 42 = 13x - 8$
h) $6q + 18 = 15q - 108$
i) $8m + 11 = 14m + 11$

Worked examples
Solve these equations.

1 $8x + 1 = 3x - 19$

$\quad -3x \qquad -3x$

$\quad 5x + 1 = -19$

$\quad -1 \quad -1$

$\quad 5x = -20$

$\quad \div 5 \quad \div 5$

$\quad x = -4$

2 $4 - 4y = 10 + 2y$

$\quad +4y \qquad +4y$

$\quad 4 = 10 + 6y$

$\quad -10 \quad -10$

$\quad -6 = 6y$

$\quad \div 6 \quad \div 6$

$\quad y = -1$

3 $-9m + 3 = 5 - 5m$

$\quad +9m \qquad +9m$

$\quad 3 = 5 + 4m$

$\quad -5 \quad -5$

$\quad -2 = 4m$

$\quad \div 4 \quad \div 4$

$\quad m = \frac{-2}{4}$

$\quad m = -\frac{1}{2}$

OR balancing the other way will give you the same answer

$-9m + 3 = 5 - 5m$

$\quad +5m \qquad +5m$

$\quad -4m + 3 = 5$

$\quad -3 \quad -3$

$\quad -4m = 2$

$\quad \div(-4) \quad \div(-4)$

$\quad m = \frac{2}{-4}$

$\quad m = -\frac{1}{2}$

---

4 Solve these equations.

a) $7x + 2 = 4x - 13$

b) $6m + 5 = 2m - 31$

c) $5p - 7 = 3p - 23$

d) $10q - 12 = 7q - 45$

e) $2w + 11 = 6w + 19$

f) $3y + 18 = 8y + 38$

g) $4n - 14 = 10n + 16$

h) $2k - 90 = 13k - 13$

i) $6m - 12 = 16m + 48$

5 Solve these equations.

a) $6 - 2x = 3x + 1$

b) $3 - 7y = 5y + 15$

c) $4 - 6w = 2w + 20$

d) $9 - 3k = 3k + 33$

e) $-4d + 2 = d + 17$

f) $-2m - 4 = 2 - 8m$

g) $-5k - 11 = 5 - 9k$

h) $-12t + 15 = 50 - 7t$

i) $-16p - 12 = 48 + 4p$

6 Solve these equations (your answer should be a fraction in each case).

a) $9x - 4 = 7x - 1$

b) $5p + 1 = p + 2$

c) $6k - 2 = 3 - 4k$

d) $11 - 2m = 6m - 9$

e) $-10p + 6 = -3 - 4p$

f) $8y + 7 = 1 - 2y$

g) $2 + 4k = -1 - 5k$

h) $-10x - 9 = 21 - 6x$

i) $7y + 6 = -2y - 15$

For the remaining worked examples we will not write the balancing steps however you should continue to do so if it helps you.

Worked example

These two triangles have the same perimeter. All lengths are in centimetres.

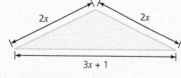 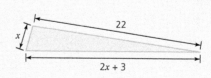

Find the dimensions of both triangles.

Perimeter of first triangle $= 2x + 2x + 3x + 1$
$\qquad\qquad\qquad\qquad = 7x + 1$

Perimeter of second triangle $= x + 2x + 3 + 22$
$\qquad\qquad\qquad\qquad\qquad = 3x + 25$

The perimeters are equal so

$7x + 1 = 3x + 25$

$4x + 1 = 25$

$\quad 4x = 24$

$\quad x = 6$

Dimensions of first triangle:

$2x = 2 \times 6 = 12$ cm

$3x + 1 = 3 \times 6 + 1$
$\qquad\quad = 19$ cm

12 cm, 12 cm and 19 cm

Dimensions of second triangle:

$x = 6$ cm

$2x + 3 = 2 \times 6 + 3$
$\qquad\quad = 15$ cm

6 cm, 15 cm and 22 cm

Exercise 71

1   The shapes in each pair have the same perimeter.
    All lengths are in centimetres and the drawings are not to scale.

    i)   Form an equation for the equal perimeters.

    ii) Solve the equation to find the marked dimensions of each shape.

    a)
    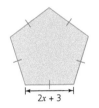
    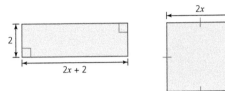

    b)
    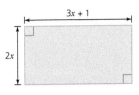
    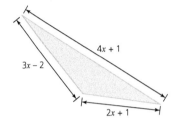

    c)
    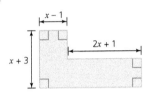

    d)

Worked example

Chelsey and Gary have three children. Their youngest child is x years old. The middle child is four years older than the youngest and the eldest is twice as old as the youngest. The sum of their ages is 32 years. Form an equation and solve it to find the age of each child.

youngest: x $x + x + 4 + 2x = 32$ youngest: 7 years old
middle: $x + 4$ $4x + 4 = 32$ middle: $7 + 4 = 11$ years old
eldest: $2x$ $4x = 28$ eldest: $2 \times 7 = 14$ years old
                                  $x = 7$

2   Abdah is y years old. His elder brother is six years older and his sister is eight years younger than him. Their ages add up to 40 years. Form an equation and solve it to find the age of each child.

3   A grape has a mass of x grams. The mass of a raspberry is two grams less than the grape and the mass of a strawberry is equivalent to five grapes. The grape, raspberry and strawberry have a combined mass of 19 g. Form an equation and solve it to find the mass of each piece of fruit.

4   In a new housing development, the first house has a garden with area x m². The garden in the second house has an area 54 m² larger than the first house and the third house has a garden area 18 m² less than the second house. Altogether, the gardens have an area of 360 m². Form an equation and solve it to find the size of each garden.

► Evaluating formulae

- Substitute using the values given for each variable.
- Evaluate, taking care to use BODMAS where necessary.

Worked example

1   For a cuboid of length l, breadth b and height h, the volume and surface area are given by:

$$V = lbh \qquad \text{and} \qquad A = 2lb + 2lh + 2bh$$

Calculate the   a)   volume of a cuboid with $l = 3\,cm$, $b = 5\,cm$ and $h = 8\,cm$

b)   surface area of a cuboid with $l = 4\,cm$, $b = 9\,cm$ and $h = 6\,cm$.

a) $V = lbh$
$= 3 \times 5 \times 8$
$= 120\,cm^3$

b) $A = 2lb + 2lh + 2bh$
$= 2 \times 4 \times 9 + 2 \times 4 \times 6 + 2 \times 9 \times 6$
$= 72 + 48 + 108$
$= 228\,cm^2$

Exercise 7J

1   Evaluate each formula for the values given.

a) $V = Ah$ when $A = 10$ and $h = 7$

b) $Q = IT$ when $I = 2$ and $T = 30$

c) $v = u + at$ when $u = 6$, $a = 2$ and $t = 3$

d) $a = \dfrac{v - u}{t}$ when $v = 15$, $u = 3$, $t = 4$

e) $c = \sqrt{a^2 + b^2}$ when $a = 3$ and $b = 4$

f) $B = \dfrac{w}{h^2}$ when $w = 84$ and $h = 2$

2   As you climb a mountain, the temperature decreases. A mountaineering group approximates the temperature, $T°C$, as you climb, using the formula

$$T = B - 2h$$

where B is the temperature at the base of the mountain, in degrees Celsius, and h is how far you have climbed up the mountain, in thousands of feet.

Calculate the temperature 3000 feet up the mountain if it is $22°C$ at the base.

3   A salesperson is paid a fixed monthly salary plus 15% commission on all sales made. This is modelled by the formula

$$W = 1200 + 0 \cdot 15s$$

where $£W$ is their wages and $£s$ is the value of sales they make.

a)   What is the salesperson's salary before they make any sales?

b)   How much will the salesperson make if they sell

i)   £300 worth of goods

ii)   £920 worth of goods?

c)   If the commission changes to 18%, what would this formula change to?

Sometimes we must balance the equation formed to find the unknown variable.

Worked examples

1 The equation of a straight line is $y = 3x + 15$.

   Find the value of

   a) y when $x = 4$ b) x when $y = 21$ c) x when $y = 90$

   a) $y = 3x + 15$ b) $y = 3x + 15$ c) $y = 3x + 15$

        $y = 3 \times 4 + 15$ $21 = 3x + 15$ $90 = 3x + 15$

           $= 27$ $6 = 3x$ $75 = 3x$

                                     $x = 2$ $25 = x$

                                                        $x = 25$

2 Given $P = 5q + 2r$, find r when $P = 16$ and $q = 2$. $P = 5q + 2r$
                                                          $16 = 5 \times 2 + 2r$
                                                          $16 = 10 + 2r$
                                                         $2r = 6$
                                                          $r = 3$

Exercise 7K

1 A straight line has equation $y = 7x + 2$. Find the value of:

   a) y when $x = 5$ b) y when $x = 11$ c) x when $y = 58$ d) x when $y = 135$.

2 Form an equation and solve it to find the letter asked for.

   a) $A = r + 5$ find r when $A = 12$

   b) $M = 3k + 2$ find k when $M = 20$

   c) $A = lb$ find b when $A = 18$ and $l = 6$

   d) $K = x + 3p$ find x when $K = 42$ and $p = 2$

3 A supplier charges £18 per pack of jotters plus a £5 delivery charge for the order. This can be represented by the formula

   $$C = 18n + 5$$

   where C is the total cost and n is the number of packs of jotters ordered.

   a) Calculate the cost of ordering 15 packs of jotters.

   b) If the jotter order costs £365, how many packs of jotters were ordered?

   The supplier has a special offer of a free pack when the order costs at least £725.

   c) Form an equation and solve it to find how many packs must be ordered to get a free pack.

4 The volume, V, of a square-based box is given by

   $$V = l^2 h$$

   where l is the length of the side of the base and h is the height of the box.

   Calculate:

   a) the height of a box with volume $100 \, cm^3$ and base length $5 \, cm$

   b) the length of the base of a box with height $10 \, cm$ and volume $40 \, cm^3$.

Check-up

1. Simplify to a single term.

   a) $x + x + x$
   b) $y + 3y$
   c) $2m + 6m$
   d) $10q - 7q$
   e) $3k + 8k - k$
   f) $4p - 2p + 7p$
   g) $6m + m - 10m$
   h) $12r - 9r - 5r$

2. Simplify:

   a) $6x + 3 + 2x$
   b) $7y + 2k - 4y$
   c) $3m + 2n + 9m + 5n$
   d) $4p + 18q - 2p + 6q$
   e) $2r + 8t + 6r - 3t$
   f) $9m + 5n - 7m - n$
   g) $6u + 4v - u - 10v$
   h) $8g - 5h + 2g + 3h$
   i) $10a - 2b - 8a - 3b$

3. Simplify:

   a) $3m + 7n + 2 + 8m - 2n$
   b) $6y + 9j - 4y + 2x - 5j$
   c) $12 + 4r + 7s - 3r + 9s$
   d) $17u - 4v + 6w - 9v + u$
   e) $10p - 3q - 5 - 2q - 8$
   f) $20x - 15y - 7x + 16k - 12y$

4. Simplify:

   a) $x^2 + 7x^2$
   b) $3y^2 + 2 + 6y^2$
   c) $9m^2 + 2m - 8m^2 + 5m$
   d) $6g^2 - 10h^2 - 4g^2 + 10j^2 + 3h^2$
   e) $2ab + 12ab$
   f) $5mn + 2n - 3mn$
   g) $8pq - 2rt + 3pq - 9rt$
   h) $4 - 3uv + 2 - 4xw + 8uv - xw$

5. Copy and complete the pyramids by adding the expressions in two adjacent boxes to obtain the expression in the box above them.

   a)
   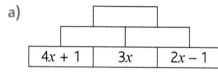

   b)
   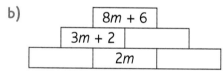

   c)
   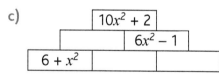

   d)
   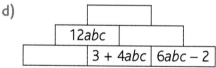

6. Substitute $p = 3$, $q = 5$ and $r = 8$ into these expressions and evaluate them.

   a) $p + 9$
   b) $2q$
   c) $3r + 1$
   d) $5p - 1$
   e) $16 - 3q$
   f) $5 + 9q - 4r$
   g) $pq - r$
   h) $rq + 6p$
   i) $5pr + 2q$
   j) $10 - 2qr$
   k) $4p^2$
   l) $(4p)^2$

7. Evaluate these expressions for $r = -2$, $s = 3$ and $t = -4$.

   a) $r + 6$
   b) $t - 10$
   c) $2s + 5r$
   d) $3t + 4s$
   e) $3r - 6s$
   f) $5t - 4r$
   g) $rs + t$
   h) $5 - st$
   i) $3rt + 4s$
   j) $9sr - 3ts$
   k) rst
   l) r^2t

8. Given $a = 2$, $b = -3$ and $c = -10$, substitute and evaluate each expression.

   a) a^2
   b) b^2
   c) $3a^2$
   d) $(3a)^2$
   e) $\dfrac{c - a}{b}$
   f) $\dfrac{2b - c}{a}$
   g) $\dfrac{4b + a}{c}$
   h) $\dfrac{ab + ac}{a + b}$

9 Solve each equation.

a) $x + 9 = 10$ b) $2y + 3 = 15$ c) $4k - 2 = 18$ d) $7w - 8 = 69$

e) $10q - 4 = 36$ f) $8p - 9 = 87$ g) $7y + 10 = 10$ h) $5m - 7 = -42$

i) $8t + 3 = -5$ j) $6v + 2 = -46$ k) $9p - 14 = -104$ l) $4m + 2 = -10$

10 Solve each equation.

a) $3x = x + 12$ b) $7k = 45 + 2k$ c) $5c = 27 - 4c$ d) $15h = 8h + 14$

e) $6p = 120 - 4p$ f) $8k + 6 = 10k$ g) $2x + 100 = 7x$ h) $5w + 24 = 8w$

i) $9x + 22 = 20x$ j) $15 - 13y = 2y$ k) $18 - 3q = 6q$ l) $48 - 7m = 5m$

11 Solve these equations.

a) $5x + 2 = x + 14$ b) $9y + 4 = 2y + 18$ c) $6m + 5 = 4m + 21$ d) $7p + 11 = 3p + 51$

e) $8h - 6 = 5h + 21$ f) $7m - 9 = 3m + 47$ g) $15k - 1 = 9k + 41$ h) $18w - 16 = 8w + 74$

12 Solve:

a) $4x + 3 = x - 9$ b) $5y + 8 = 3y - 20$ c) $7m + 9 = 3m - 23$ d) $6q + 14 = 2q - 50$

e) $8w - 4 = 2w - 46$ f) $10k - 2 = 3k - 65$ g) $5j - 28 = 7j - 6$ h) $3r - 140 = 13r - 20$

13 The two shapes have the same perimeter. All lengths are in centimetres and the drawings are not to scale.

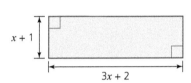
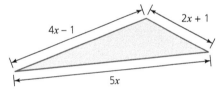

a) Form an equation for the equal perimeters.

b) Solve the equation to find the dimensions of each shape.

14 Evaluate each formula for the values given.

a) $V = lb^2$ Evaluate V when $l = 8$ and $b = 5$.

b) $A = \frac{1}{2}$ of $h \times (a + b)$ Evaluate A when $h = 6$, $a = 4$ and $b = 9$.

c) $A = 2l + 4lh$ Evaluate A when $l = 3$ and $h = 10$.

15 Form an equation and solve it to find the letter asked for.

a) $W = 3n + 4$ Find n when $W = 52$.

b) $H = 10 + 4k$ Find k when $H = 70$.

16 The total weight, W kg, of a lorry can be approximated using the formula

$$W = 20b + 8000$$

where b is the number of bags that can be filled with the gravel in the lorry.

a) What is the weight of an empty lorry?

b) Calculate the total weight of a lorry carrying enough gravel for 70 bags.

The maximum weight limit on a weak country bridge is 14 000 kg.

c) What is the maximum number of bags that can be filled from a lorry which can safely cross the bridge?

▶ Best value

Shops usually sell different sizes of popular items. It is important to know how to check which are the best value for money.

- One way to compare prices is to use direct proportion.
- To scale quantities in direct proportion, multiply or divide both sides by the same amount.

Worked example Decide which size offers the better value for money.

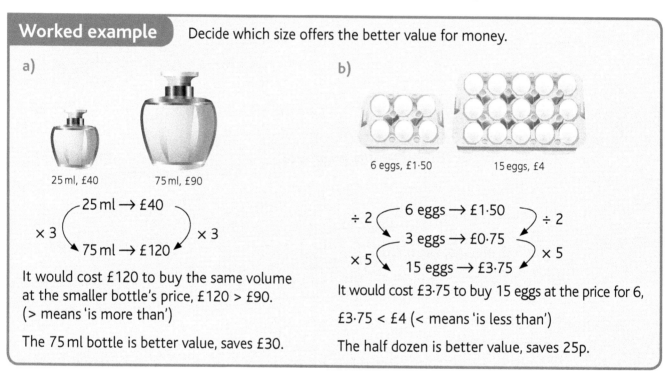

a)

25 ml, £40   75 ml, £90

$$25\,ml \rightarrow £40$$
$$\times 3 \qquad \qquad \times 3$$
$$75\,ml \rightarrow £120$$

It would cost £120 to buy the same volume at the smaller bottle's price, £120 > £90. (> means 'is more than')

The 75 ml bottle is better value, saves £30.

b)

6 eggs, £1·50   15 eggs, £4

$$\div 2 \qquad 6\,eggs \rightarrow £1\cdot50 \qquad \div 2$$
$$3\,eggs \rightarrow £0\cdot75$$
$$\times 5 \qquad 15\,eggs \rightarrow £3\cdot75 \qquad \times 5$$

It would cost £3·75 to buy 15 eggs at the price for 6,

£3·75 < £4 (< means 'is less than')

The half dozen is better value, saves 25p.

Exercise 8A

1   Decide which deal offers the best value for money.

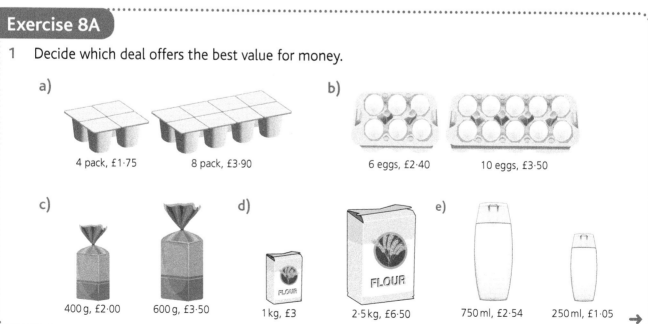

a)

4 pack, £1·75   8 pack, £3·90

b)

6 eggs, £2·40   10 eggs, £3·50

c)

400 g, £2·00   600 g, £3·50

d)

1 kg, £3   2·5 kg, £6·50

e)

750 ml, £2·54   250 ml, £1·05

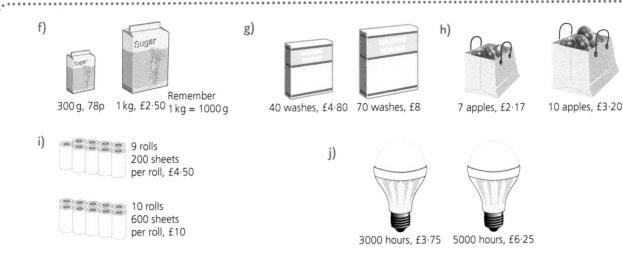

f) 300 g, 78p   1 kg, £2·50   Remember 1 kg = 1000 g

g) 40 washes, £4·80   70 washes, £8

h) 7 apples, £2·17   10 apples, £3·20

i) 9 rolls 200 sheets per roll, £4·50

10 rolls 600 sheets per roll, £10

j) 3000 hours, £3·75   5000 hours, £6·25

2   Cathal spots these deals for the same items in different sizes.

He is unsure if the deals offer good value or not. Do you think they do? Explain why.

Item	Great deal?
a) Tinned tomatoes 95p a tin	4-pack of tins, £4·80
b) Bread £1·80 a loaf	3 loaves, £5
c) Strawberries 200 g, £1·70	Big value pack 500 g, £5
d) Tea bags 80 bags, £2	Savers pack 200 bags, £5
e) Cat food, single pouch 72p	Mega pack 90 pouches, £54

3   Brodie usually buys a large tin of beans.

He notices a smaller tin on special offer.

Which option is better value?

400 g tin, £2·30   200 g tin, £1·80
buy one get one free

4   Yuwen usually buys a large bottle of fabric softener.

She notices a smaller one on special offer.

Which option is better value?

500 ml, £4·50
buy three for the price of two

1 litre, £7

5   Kayleigh likes protein bars and pays £1·28 for one from her local shop. She can buy a box of 40 bars for £38·40.

How much could Kayleigh save, per bar, by buying in bulk?

6   Sadie is buying balls of wool.

There is a **Buy one get two free** deal on the wool.

One ball of wool normally costs £7·20.

Sadie stocks up and buys 12 balls of wool.

a) How much will Sadie pay for 12 balls of wool?

b) Calculate how much she has saved in total.

▶ Budgeting

Planning your spending to match your earning is called budgeting. Every year both the Scottish and UK governments set out their plans to balance income and expenditure for the country. Governments, global businesses and charities balance their budgets by using the same mathematics we do.

- Add up the total available to spend.
- Add up how much you are planning to spend.
- Find the difference.
- Adjust your plans to avoid spending more than you have.

Incoming

Outgoing

Balance = Incoming – Outgoing

Worked example   Mya has a budget of £190 to host a party. Her spending plan is shown.

a)   Complete the budget.

b)   What advice would you give her?

a)

125·00
46·90
52·85
+45·00
269·75

DJ	£125		
Buffet	£46·90		
Decorations	£52·85	Incoming	£190
Hall Hire	£45	Outgoing	£269·75
Total	£269·75	Balance	−£79·75

Balance = Incoming – Outgoing
= 190 – 269·75
= −£79·75

b)   The balance is negative. Mya is spending nearly £80 too much.

She could spend less on decorations and look for a less expensive DJ.

Exercise 8B

1   Pat is planning a children's party. She has £200 to spend.

   a)   Copy and complete the budget.

   b)   Does Pat have enough money for her plans?

Entertainer	£92·50		
Buffet	£55		
Decorations	£30	Incoming	£200
Cake	£14·95	Outgoing	
Total		Balance	

2   William is an apprentice commis chef. He lives with his dad and his net (take home) pay is £970 a month.

   a)   Copy and complete William's monthly budget.

   b)   How much will William have left to spend per week, in a 4-week month?
   (Divide the balance by 4).

   c)   Do you think William should spend all of his money every month?

Rent, food and bills	£425		
Mobile	£50	Incoming	£970
Travel Pass	£55	Outgoing	
Total		Balance	

→

3 Juliette is in her first year of an apprenticeship to become a power network worker.

Her net pay is £869 a month.

a) Copy and complete Juliette's budget.

Rent, food and bills	£250		
Mobile	£64		
Travel Pass	£35	Incoming	
Savings	£125	Outgoing	
Total		Balance	

b) How much will Juliette have left to spend per week, in a 5-week month?

4 Danni is in her third year of a straight-from-school accountancy apprenticeship. She is gaining qualifications and working; her take home pay is £1300 a month.

Rent, food and bills	£540		
Mobile	£21		
Car	£230		
Gym	£29·50		
Clothes	£260	Incoming	
Savings	£65	Outgoing	
Total		Balance	

a) Copy and complete Danni's monthly budget.

b) Danni thinks she spends half her money on essentials. Do you agree?

c) What fraction of her monthly wage is Danni spending on clothes? Simplify your answer.

d) What percentage of her money is Danni saving each month?

5 Lewis works as a customer service advisor. His budget for a month is shown below.

Rent, food and bills	£575		
Credit card payment	£70		
Car payment	£164.50		
Savings	£200	Incoming	£1120
Gig tickets		Outgoing	
Total		Balance	£65.50

Copy and complete Lewis's monthly budget.

▶ Using technology for budgeting

- We can set up a **spreadsheet**, using formulae to carry out calculations automatically.

If you have access to technology, try these questions for yourself too.

6 Mark works as an office administrator. His net pay is £1125 per month and he shares a flat with a friend. The spreadsheet shows his monthly budget.

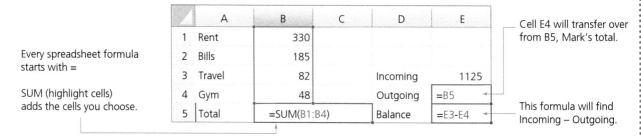

Every spreadsheet formula starts with =

SUM (highlight cells) adds the cells you choose.

	A	B	C	D	E
1	Rent	330			
2	Bills	185			
3	Travel	82		Incoming	1125
4	Gym	48		Outgoing	=B5
5	Total	=SUM(B1:B4)		Balance	=E3-E4

Cell E4 will transfer over from B5, Mark's total.

This formula will find Incoming – Outgoing.

a) What value will be in cell B5? This is the total of Mark's spending.

b) What value will be in cell E5? This is his monthly balance.

c) Is there anything that you think Mark has forgotten from his budget?

7 Kesia takes the maximum student loan of £4740 a year and earns £320 a month working part-time. The spreadsheet shows her monthly budget.

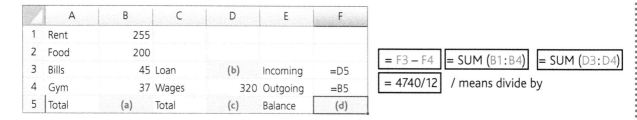

	A	B	C	D	E	F
1	Rent	255				
2	Food	200				
3	Bills	45	Loan	(b)	Incoming	=D5
4	Gym	37	Wages	320	Outgoing	=B5
5	Total	(a)	Total	(c)	Balance	(d)

= F3 – F4	= SUM (B1:B4)	= SUM (D3:D4)
= 4740/12	/ means divide by	

Match a), b), c) and d) from the spreadsheet with the formulae given.

e) Complete Kesia's budget. How much does she have left to spend each month?

8 A group of S3 pupils are running a wellbeing day. They use a spreadsheet to keep track of costs.

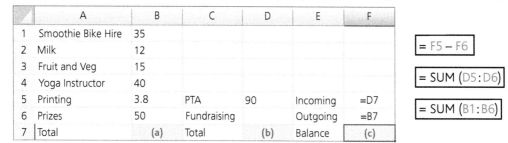

	A	B	C	D	E	F
1	Smoothie Bike Hire	35				
2	Milk	12				
3	Fruit and Veg	15				
4	Yoga Instructor	40				
5	Printing	3.8	PTA	90	Incoming	=D7
6	Prizes	50	Fundraising		Outgoing	=B7
7	Total	(a)	Total	(b)	Balance	(c)

= F5 – F6
= SUM (D5:D6)
= SUM (B1:B6)

Match a), b) and c) from the spreadsheet with the formulae given.

d) How much do the pupils need to fundraise (cell D6) to balance the budget above?

e) The pupils convince a local shop to donate the milk and fruit and veg and they fundraise £85. Calculate their new balance.

▶ Bank statements

It is very important to keep track of your money by checking your balance (how much money is in the account).

To calculate the balance:

- add or subtract each transaction from the previous balance.
- the table explains some common transactions.

CRE	Credit	Money has been paid into your account.
		For example: wages, benefits, shop refunds, cash deposits.
DBT	Debit	Money has been paid out of your account.
		For example: withdrawing cash and spending by using a debit card.
DR	Account overdrawn	Your balance is below zero.
D/D	Direct debit	A regular payment of a fixed or variable amount that you have agreed to. For example: energy bills, credit card payments, hire purchase, gym membership.
S/O	Standing order	A regular payment of a fixed amount that you have set up.
		For example: paying into a savings account, charitable giving, paying rent.
INT	Interest	Interest may be paid on a positive balance.
		Interest may be charged on a negative balance.
CHG	Charge	Bank charge. For example: being overdrawn, withdrawing foreign currency.

Worked example

Complete the balance column on this bank statement.

						−£100 DR
		Balance brought forward				
25 Jun	CRE	Salary	£1100			£1000
25 Jun	DBT	Card transaction		£75·99		£924·01
26 Jun	CHG	Overdraft		£12		£912·01
26 Jun	DBT	Card transaction		£152·51		£759·50
30 Jun	S/O	Rent		£475		£284·50
2 Aug	CRE	Cash paid in	£80			£364·50

Payments in are listed in this column.

Payments out are listed in this column.

This column shows the balance after each transaction.

- Set down a sum or use a calculator to find the balance after each transaction.

Exercise 8C

Complete the highlighted balance column of the bank statements shown.

1

		Balance brought forward			£985
21 Dec	DBT	Card transaction		£15	
21 Dec	DBT	Card transaction		£54	
24 Dec	DBT	Card transaction		£120	
24 Dec	CRE	Refund	£25		
29 Dec	D/D	Car payment		£237	

2

		Balance brought forward			£72·10
13 Feb	CRE	Department work and pensions	£85		
14 Feb	D/D	Mobile contract		£22	
14 Feb	DBT	Cash withdrawal		£50	
14 Feb	CRE	Salary	£965·09		
17 Feb	D/D	Gas bill		£41	
20 Feb	D/D	Store card payment		£42	

3

		Balance brought forward			£250
4 Aug	CRE	Salary	£1200		
5 Aug	DBT	Card transaction		£64·25	
9 Aug	D/D	Loan payment		£128·20	
10 Aug	DBT	Card transaction		£32·71	
11 Aug	D/D	Mortgage		£578	
14 Aug	D/D	Council tax		£141	

4

		Balance brought forward			−£500 DR
1 Sep	CRE	Salary	£1105		
2 Sep	DBT	Card transaction		£75·99	
4 Sep	CHG	Overdraft		£12	
4 Sep	DBT	Card transaction		£152·51	
7 Sep	S/O	Rent		£475	
8 Sep	D/D	Council tax		£153	
10 Sep	D/D	Electricity bill		£37	

Complete the highlighted boxes of the bank statement shown.

5

		Balance brought forward			−£ DR
1 May	DBT	Loan payment		£118	−£281 DR
1 May	DBT	Rent		£315	
1 May	CRE	Salary			£1645
2 May	D/D	Council tax			£1521·46
3 May	D/D	Gas and electricity		£80·29	
3 May	CRE	Salary back pay			
10 May	DBT	Card payment		£37	£2784·29

▶ Personal finance: contracts and services

When deciding how to spend your money you will often have to choose between different payment options.

● Work out the overall cost of any contracts. Be certain you can afford to pay, before making a commitment.

● Find out what your options are. There is often a deal to be found.

● Careful planning almost always saves money. Make sure you are not wasting your hard-earned cash.

Worked examples

1   Mitchell is looking at tram tickets and plans to take 4 journeys today.

How much will he save by switching from single tickets to an all-day ticket?

Single	£1·70
All day	£4

$$
\begin{array}{r} 1{\cdot}70 \\ \times\ _24 \\ \hline 6{\cdot}80 \end{array}
\qquad
\begin{array}{r} 6{\cdot}80 \\ -\ 4{\cdot}00 \\ \hline £2{\cdot}80 \end{array}
$$

Mitchell will save £2·80 by buying an all-day ticket.

2   Ava is looking at two options for broadband. Both providers are offering the same speed.

a)   Calculate the total cost of each contract.

b)   Which contract would you recommend?

Scotnet	ConnectU
£30 a month	Just £5 a month*
£37 set up fee	12 month contract
12 month contract	*first 3 months only, regular price £40 a month

a)   Scotnet

$$30 \times 12 = 360$$
$$\begin{array}{r} +\ \ 37 \\ \hline £397 \end{array}$$

ConnectU

$$5 \times 3 = \quad 15$$
$$\begin{array}{r} 40 \times 9 = +360 \\ \hline £375 \end{array}$$

b) $$\begin{array}{r} 397 \\ -375 \\ \hline £22 \end{array}$$

Ava will save £22 by choosing ConnectU.

Exercise 8D

1   Morag is looking at the options for her to get to work. She works 5 days a week.

Option	Cost per day	Pros	Cons
Bus	£3·20	every 10 minutes	takes longer, busy traffic
Car	£5	home earlier	busy traffic, stressful
Train	£2·60	free wifi	only one train an hour

a)   What is the difference between the most expensive and least expensive options for Morag to get to work for a week?

b)   Which option would you recommend?

→

2 Mominah is looking at the following choices for her bus ticket.

Single journey	£1·85
All day	£5·50
Weekly	£17·50
4 week	£52
10 week	£110

Mominah takes 2 buses to work, that is 4 journeys per day. She works for 5 days a week.

a) How much will Mominah save in 1 day by switching from single journey tickets to an 'all-day' ticket?

b) How much would she save in a week by switching from 'all-day' tickets to a weekly ticket?

c) How much would she save per week by switching from a weekly to a 4-week ticket?

Mominah's mum lends her the money to buy a 10-week ticket.

d) If Mominah only uses her ticket to go to work how much will she pay, per journey, using the 10-week ticket?

3 Miss Finlay buys a football season ticket for £495. The ticket allows entry into 21 home games which normally cost £25 a ticket and 3 special games which usually cost £45 a ticket.

a) Calculate Miss Finlay's total possible saving if she attends every match.

b) She misses all the £45 games and only attends 9 regular games. Calculate how much she could have saved by not buying the season ticket.

4 Stephen is looking at choices for an internet TV subscription site. The table shows monthly payments.

Number of screens	Cancel anytime	6 month membership*	18 month membership*
1	£8·99	£7·99	£4·99
2	£9·99	£8·99	£5·99
Unlimited	£12	£11	£8
*Note, no cancellation possible during term of contract			

a) Calculate the total cost of an 18-month membership for 2 screens.

b) Calculate the total cost of a 6-month membership for 1 screen.

c) How much more is it per month to switch from 1 screen to an unlimited number of screens on the cancel anytime plan?

d) Stephen shares a flat with 3 friends. He has not tried this subscription site before. Which option would you recommend?

5   Marian is looking at broadband deals. She is comparing 3 deals which are all 12-month contracts with equal speeds and unlimited downloads.

Plus Link	Get Online	Communicare
£4 per month! * *first 3 months only, then £42 a month	£29 a month + one off £72 activation fee	£9·99 a month + £23 monthly line rental

   a)   Calculate the amount you will pay over the year (annual cost) on the Plus Link deal.

   b)   Calculate the annual cost of the Get Online deal.

   c)   Calculate the annual cost of the Communicare deal.

   d)   Which deal would you recommend?

6   The table shows the payment options for over 16s at a leisure centre.

Fitness sessions	Fitness combos	Fitness for less
Swim £5·60	Swim & skate £9·50	£30 per month min 12-month contract
Skate £6·50 + £3 hire	Swim & sauna £9·50	£300 Annual Fee
Sauna £7·50	Skate & sauna £9·50	

   a)   Rachel swims and skates on a Saturday, using her own skates. How much could she save with a combo ticket?

   b)   Which combo ticket is the best value for money? Explain your answer clearly.

   c)   Rachel only uses the centre at the weekend. She buys a Swim & skate combo every Saturday and a Sauna ticket on a Sunday. How much could Rachel save, in a 4-week month, by switching to the Fitness for less contract?

   d)   Calculate the total cost of the 12-month Fitness for less contract.

   e)   Which option would you recommend?

7   Beth is looking at mobile phone options. She uses between 3 and 5 GB of mobile data a month. Beth mainly uses a messaging app to contact her friends and uses very few minutes and texts. She already owns a smartphone and is looking at SIM-only deals.

A £9 a month 12-month contract 1000 minutes, 1000 texts 4GB data* *excess data charged £1 per MB	B £15 a month 18-month contract 500 minutes, 500 texts 4·5GB data* *excess data charged at £1 a day	C £23 a month 18-month contract 800 minutes, 800 texts Unlimited data

   a)   Which contract would you recommend for Beth? Explain your answer clearly.

   b)   Calculate the total which Beth would pay over the full term of contract C.

   c)   Calculate the total which Beth would pay over the full term of contract B, assuming she did not exceed the data limit.

8   Ms Grant is renovating a bungalow.
    She looks at some choices for the heating system.

   a)   Which option would you recommend?

   b)   How many years will it be until the ground source heat pump works out cheaper?

Gas central heating	Ground source heat pump
Fossil fuel	Sustainable energy
Installation £1850	Installation £5100
Annual cost £1200	Annual cost £90

► Personal finance: saving money

You can save money by making smart choices about goods and services. Saving money also means putting money aside and not spending it. It is a good idea to use a savings account that pays interest on your savings.

Different banks and accounts have different interest rates. This is the amount the bank pays you to leave your money with them. For example, an interest rate of 2% p.a. (per annum or each year) means the bank will pay you 2% of your balance and add it to your savings after one year.

In this exercise we shall assume the savers deposit their money in the bank and do not spend any of it for 1 year.

Worked examples

1 Margaret saves £4000 in an account which pays 3% interest p.a. Calculate her balance after 1 year.

1% = 40	Interest = £120	4000	Balance after 1 year = £4120
×3		+ 120	
3% = 120		4120	

2 Bogdan saves £300 in an account which pays 1·5% interest p.a. Calculate Bogdan's balance after 1 year.

1% = 3	Interest = £4·50	300·00	Balance after 1 year = £304·50
+0·5% = 1·50		+ 4·50	
1·5% = 4·50		304·50	

Exercise 8E

1 Gordon has £500 and puts it in a savings account paying 2% p.a. Calculate his balance after 1 year.

2 Adriana saves £180 in a savings account paying 4% p.a. Calculate her balance after 1 year.

3 Kwai receives £60 for his birthday and saves it in an account paying 5% p.a. Calculate his balance after 1 year.

4 Natalie puts £780 in a savings account paying 3% p.a. Calculate her balance after 1 year.

5 Calah saves £240 in an account paying 1·5% p.a. Calculate her balance after 1 year.

6 Carolanne saves £790 in a savings account. Her balance after 1 year is £809·75. She thinks the interest rate is 2% p.a. but her friend Maureen thinks it is 2·5%. Who is correct? Show working to explain your answer.

7 Anne has £6400 to save. She sees two savings account offers.

a) Which savings account would you recommend?

b) How much more interest will Anne earn by choosing the better deal?

Savings interest rate	
Family Trust	1·25% p.a.
Bonny Life	3% p.a.

➜

8   Iona, Maisie and Richard win 1st, 2nd and 3rd prizes in a maths competition. They each save their money in a different one of the banks shown. The balances of their accounts after 1 year are shown.

	Prize	Balance after 1 year
Iona	£250	£251·25
Maisie	£245	£252·35
Richard	£240	£253·20

Savings interest rate	
Scots Savers	5·5% p.a.
Alba Bank	0·5% p.a.
Society Bank	3% p.a.

Which bank did each person use? Show working to explain your answer.

9   Sally saves £19 a week from her Saturday job. How much will Sally have after 20 weeks?

10   Emma-Jane is saving up to buy a new bathroom. Her savings target is £900. Emma-Jane saves £50 a week. How long will it be until she can buy her new bathroom?

11   Hardeep is saving to buy a holiday. His savings target is £1500. Hardeep manages to save £250 a month. How long will it be until he can afford the holiday?

12   Raymond is using a supermarket scheme to save up for Christmas. He buys 'stamps' for £5 each. When he fills a card with 12 stamps he gets another stamp free. The stamps are then swapped for vouchers to spend in the supermarket.

   a)   Raymond fills 3 cards. How much will he have to spend at Christmas?

   b)   What are the benefits of this scheme? Are there any downsides?

13   Jenny is given £450 for her birthday and she invests it in a savings account. She does not spend any of it. One year later her balance is £461·25. Find the interest rate on her savings account.

▶ Personal finance: borrowing money

There are lots of ways to borrow money. For example: loans, credit cards, overdrafts and mortgages.

Shops often offer hire purchase or finance contracts which allow you to pay for larger purchases in instalments over time. These are financial products and are ways to borrow money.

When you borrow money you almost always pay interest. This means you will pay back more than you receive.

All financial products must clearly display an APR (Annual percentage rate). This is the total cost of borrowing the money for a year as a percentage of the amount borrowed.

Lower APR, lower charges ——→ higher APR, higher charges

To compare the prices of financial products:

● Calculate the total you will have to pay back, subtract the cash amount.

● Look for a lower APR; this gives an indication of how much interest you will pay.

Borrowing money can be a useful part of a well-planned financial life. It can also lead to serious money problems. It is important to understand the process of borrowing and to think through the long-term consequences of borrowing money.

Worked example   Max buys a tablet computer using a **finance agreement** or **hire purchase** deal.

Calculate how much more the hire purchase deal will cost.

Cash price £610

Hire purchase

£20 a month for
48 months.

Total repayments	960
= 48 × 20	−610
= 960	350

Paying by instalments costs £350
more than buying the computer
outright.

Exercise 8F

1   Calculate how much more each hire purchase or finance agreement will cost.

   a) Rory is buying a cooker.

   Hire purchase

   £40 a month for
   24 months

   Cash Price £800

   b) Caroline is buying a used car.

   Hire purchase

   £100 a month for
   36 months

   Cash Price £2800

   c) Ronan is buying a washing machine.

   Hire purchase

   £19 a month for
   24 months

   Cash price £395

   d) Francis is buying new furniture.

   Cash price £1500

   Hire purchase

   £38 a month for
   48 months

   e) Graham is buying a new kitchen.

   Cash Price £3000   Finance agreement

   £90 a month for 5 years

   f) Malik is buying a new bathroom.

   Cash Price £7200   Finance agreement

   £180 a month for 48 months

2   List these loan companies from the least to most expensive.

Company	Happy Loanz	Smile Finance	2dayPay
APR	28%	16·5%	328·9%

3   Antoni is borrowing £500 to pay for car repairs. He looks at some options.

Daypay Bank APR 20·9%	Fife Credit Union APR 1·8%	Bank of Edinburgh APR 9·9%
60 monthly payments of £13·01	12 monthly payments of £42·07	24 monthly payments of £22·95

   a)   Which loan would you choose based on the APR?

   b)   Calculate the total to be repaid for each loan.

   c)   What is the difference in cost between the most and least expensive finance options?

► Foreign exchange

To convert pounds sterling to a foreign currency we need to use the **exchange rate**.

This tells you what £1 buys in each currency.

- £ → foreign currency

Multiply by the exchange rate.

	We sell	Rounding
Euro (€)	€1·14	nearest cent (2dp)
US dollar (US$)	$1·27	nearest cent (2dp)
Indian rupee (₹)	88·19 ₹	nearest rupee
Chinese yuan (¥ Chn)	8·75 ¥	nearest fen (2dp)
Russian rouble (₽)	81·59 ₽	nearest kopek (2dp)
Japanese yen (¥ Jpn)	139·69 ¥	nearest yen

Worked examples 1 Use the exchange rates above and a calculator to convert:

a) £600 → Chinese yuan
    600 × 8·75 = 5250 ¥ CHN

b) £450 → Japanese yen
    450 × 139·69 = 62860·5
    = 62861 ¥ Jpn

2 An auction house sells a rare whisky to a Russian collector.
   The sale price is £43 000. Calculate the price in Russian roubles.
   43000 × 81·59 = 350 8370 ₽

Exercise 8G

You may use a calculator for this exercise.

1 What is £100 worth in:
   a) Chinese yuan   b) euros   c) Indian rupees
   d) US dollars   e) Russian roubles f) Japanese yen?

2 What is £3000 worth in:
   a) Indian rupees   b) Japanese yen   c) Russian roubles
   d) US dollars   e) Chinese yuan   f) euros?

3 Convert to the currency shown.
   a) £900 → Chinese yuan   b) £600 → euros   c) £700 → Indian rupees
   d) £2000 → Japanese yen   e) £180 → Russian rouble f) £50 → US dollars
   g) £4000 → Chinese yuan   h) £59·99 → euros   i) £251·60 → Indian rupee

4 Convert:
   a) £890 → €   b) £625 → $ c) £4350 → ₽
   d) £710 → ¥ Jpn e) £19450 → ¥ Chn f) £595 → ₹

5 Mr Carling sells cashmere gloves to a distributor in China. The total value of the goods is £160 000. Calculate the price in Chinese yuan.

6 Maria has developed a languages app which she is selling to a company in California. The sale price is £775 000. Calculate the sale price in US dollars.

7 A Scottish company provides mobile electricity generators to Chennai in India. The price of hiring the generators is £27 000 per month. Calculate the price in Indian rupees.

To convert back to pounds sterling divide by the exchange rate.

- Foreign currency → £

  Divide by the exchange rate

Euro (€)	€1·14
US dollar (US$)	$1·27
Indian rupee (₹)	88·19 ₹
Chinese yuan (¥ Chn)	8·75 ¥
Russian rouble (₽)	81·59 ₽
Japanese yen (¥ Jpn)	139·69 ¥

Worked example Find the value of 20 000 ₽ Russian roubles in pounds sterling.

$20000 \div 81·59 = 245·128$ Use a calculator

$= £245·13$ Round to nearest penny

8 Write each amount in pounds sterling:
  a) 5000 Chinese yuan
  b) 890 Russian roubles ₽
  c) 400 US dollars
  d) €5000
  e) 2000 ₹
  f) 900 ¥ Jpn
  g) 30 000 ₽
  h) $1 000 000
  i) 1 000 000 ₹
  j) €368
  k) 4581 ¥ Chn
  l) 799·99 US dollars

9 A one-bedroom flat is for sale near Berlin for €180 000. Calculate the price in pounds sterling.

10 A one-bedroom flat is for sale near Tokyo for 48 800 000 ¥ Jpn. Calculate the price in pounds sterling.

11 Nicky has ordered 81 000 ¥ of gymnastics equipment from a factory in China. Calculate the price in pounds sterling.

12 A watch maker in Orkney buys 200 g of the precious metal palladium from a supplier in Mumbai, India. The cost is 793 710 ₹.
  a) Calculate the cost of the palladium in pounds sterling.
  b) Calculate the cost of one gram of palladium in pounds sterling.

× by rate

£ UK   Foreign Currency

÷ by rate

- £ → foreign currency, multiply by the rate
- Foreign currency → £, divide by the rate

Worked example

Brian and Jeanne are going to Florida. They have saved up £2000 for the trip. They spend $1980 in Florida. They change their remaining money to pounds sterling on their return. How much money will they receive?

$2000 \times 1·27$ Multiply to $2540 - 1980$ Subtract the $560 \div 1·27 = 440·944$ Divide to

$= \$2540$ convert to $. $= \$560$ spending. $= £440·94$ convert to £.

For questions 13–16, continue to use the exchange rates at the top of page 128.

13 Kirsten is travelling to Shanghai. She changes £900 into Chinese yuan and spends 6000 ¥ in Shanghai. She converts her remaining money to pounds sterling on her return. How much money will she receive?

14 Evie travels to Paris and Barcelona. She saves £1500 for her trip. Evie spends €850 in Paris and €801 in Barcelona. She converts her remaining money to pounds sterling on her return. How much will she receive?

15 Elisabeth and Patricia are planning a trip to Rome.

    a) The table shows their total expenses for the two of them.

       Elisabeth and Patricia have saved up £900 for the trip.

       Convert their savings to euros, this is the 'Incoming' figure.

Accommodation	€565		
Meals	€178	Incoming	€...
Roma pass (travel + museums)	€77	Outgoing	€...
Total		Balance	€...

    b) Complete the budget table. Calculate how many euros they will have left to spend as they like.

    c) How much is their balance in pounds sterling?

16 A phone manufacturer sells the same models of smartphone at different retail prices around the world.

    China 5388¥   USA $700   UK £539   France €636

    a) Convert each price to pounds sterling.

    b) In which currency is the phone most expensive?

    c) In which currency is the phone least expensive?

Banks and shops that sell currency buy it back at a less favourable rate.

    ● £ → Foreign currency, multiply by the 'we sell' rate.

    ● Foreign currency → £, divide by the 'we buy' rate.

17 a) Frank plans a trip to Paris and exchanges £800. However, he doesn't go, so he changes his money back to pounds sterling.

       How much money did Frank lose by doing this?

	We sell	We buy
Euro	1·14	1·36
Dollar	1·27	1·4

    b) Tara converts £21 000 to US dollars for a business deal. The deal falls through and she changes the money back.

       How much money did Tara lose by doing this?

18 Mr McCall sees these rates for buying the €.

    a) Where do you think Mr McCall should change his money?

Supermarket	Bank	Airport
£1 = €1·19	£1 = €1·21	£1 = €1·04

    b) Mr McCall has £2500 to change. Calculate the difference in what he will receive using the best and worst rates. Give your answer in euros.

Check-up 👍

1. Which deal is the better value? Show working to explain your answer.

   a)

   500 g, £2·65   4 kg, £14·50

   b)

   300 ml, £11·25   700 ml £27

2. Joseph usually buys a big bottle of Ecowash.

   He notices a smaller bottle on special offer. Which is better value? Show working to explain your answer.

   20 washes, £5   80 washes, £9
   buy one get one free

3. You have a total of £150 to spend on a party. There are 25 guests coming.

   Copy and complete the table to plan your budget.

   Choose prices from the options above.

Food	Home-made buffet £3 per person	Shop-bought £5·50 per person
Decorations	Fancy, £45 total	Regular, £20 total
Entertainment	Friend, £10	DJ £55

Decorations			
Food		Incoming	£150
Entertainment		Outgoing	
Total		Balance	

4. Reigan is looking at some options to pay for driving lessons.

   a) Is a 2-hour lesson better value than a 1-hour lesson? Explain your answer clearly.

   b) How much could Reigan save per lesson by switching from one-hour lessons to buying a block of 10?

   c) How much could Reigan save per lesson by switching from buying a block of 10 to a block of 15?

   d) Reigan has not had any lessons with this driving instructor yet. Which option would you recommend to start with?

One-hour lesson	£28
Two-hour lesson	£52
Block of 10 one-hour lessons	£240
Block of 15 one-hour lessons	£340

5. Calculate how much more each finance deal will cost.

   a) Mrs Murray is buying a sofa

   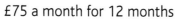

   Hire purchase

   £75 a month for 12 months

   Cash price £849

   b) Mr Taylor is buying a used car

   Finance agreement

   £108 a month for 4 years.

   Cash price £3500

6 Louis has £820. He saves it in an account that pays 2% p.a. What will his balance be after 1 year if he doesn't spend any of his money?

7 Matilda is looking at two finance options to buy a preowned car. The car costs £4000 to buy outright.

EZ Finance APR 35%	Loans4cars APR 5%
£120 a month for 120 months	£175 a month for 24 months

a) Which option would you choose, based on the APR?

b) Calculate the total repayments with Loans4cars.

c) Calculate the total repayments with EZ Finance.

d) How many years will Matilda make payments for on the EZ Finance deal?

e) Matilda only has £120 a month to spend on the car. Do you think she should take the EZ Finance deal?

8 Complete the balance column highlighted on the bank statement.

		Balance brought forward			£590·50
23 Feb	CRE	Salary	£1340		
24 Feb	D/D	Rent		£525	
25 Feb	DBT	Cash withdrawal		£40	
28 Feb	S/O	Savings		£75	
28 Feb	CRE	Credit union prize draw	£160		
2 Mar	D/D	Electricity		£52	

9 The table shows the exchange rates for £1.

Use the rates to exchange each amount as shown.

You may use a calculator for these questions

Norwegian krone	11·05 kr
South Korean won	1509 ₩
Hong Kong dollar	9·92 HK$
Euro	€1·14

a) £10 → HK$

b) £100 → HK$

c) £500 → HK$

d) £1000 → ₩

e) £900 → ₩

f) £6000 → kr

g) 1984 HK$ → £

h) 15090 ₩ → £

10 A company sells the same tablet computers at different prices around the world.

9480 kr   7722 HK$ 750 000 ₩

In which currency is the computer most expensive? Use the exchange rates shown above.

11 Marcella and Leon are planning a visit to Warsaw in Poland.

The exchange rate is £1 → 4·87 zł

a) Copy the table. It shows their total expenses for two.

Marcella and Leon have saved up £500 for the trip.

Convert their savings to Polish zloty, this is the 'Incoming' figure.

Accommodation	1125 Zł			
Meals	450 Zł	Incoming		Zł
Travel passes	39 Zł	Outgoing		Zł
Total		Balance		Zł

b) Complete the budget table. Calculate how many zloty they will have left to spend as they like.

c) How much is their balance in pounds sterling?

9 Patterns and relationships

▶ Sequences

A **sequence** is a set of numbers linked by a rule. A particular number in a sequence is a **term**. In the sequence 4, 7, 10, 13, 16, 19 ... the first term is 4, the second term is 7, the third term is 10 and so on.

To describe a sequence, state the first term and write down the rule that connects one term to the next.

Worked examples

1   These sequences involve addition or subtraction. Write down the next three terms and write down a rule that describes the sequence.

   a)   12, 19, 26, 33, 40, 47, 54 ...

   Start at 12 and add 7 each time.

   b)   45, 41, 37, 33, 29, 25, 21 ...

   Start at 45 and subtract 4 each time.

2   These sequences involve multiplication or division. Write down the next three terms and write down a rule that describes the sequence.

   a)   4, 8, 16, 32, 64, 128, 256 ...

   Start at 4 and multiply by 2 each time.

   b)   2187, 729, 243, 81, 27, 9, 3 ...

   Start at 2187 and divide by 3 each time.

3   The first term of a sequence is 12 and the rule is 'add 7 each time'. What is the fourth term in the sequence?

   2nd term: 12 + 7 = 19   3rd term: 19 + 7 = 26   4th term: 26 + 7 = 33.

Exercise 9A

1   These sequences involve addition or subtraction. For each one, write the next three terms and the rule.

   a)   2, 5, 8, 11 ...   b)   9, 15, 21, 27 ...   c)   1, 14, 27, 40 ...   d)   25, 23, 21, 19 ...
   e)   63, 55, 47, 39 ...   f)   86, 72, 58 ...   g)   14, 8, 2, –4 ...   h)   –19, –7, 5 ...
   i)   113, 94, 75 ...   j)   –3, –11, –19 ...   k)   13, 27, 41 ...   l)   7, 8·3, 9·6 ...

2   These sequences involve multiplication or division. For each one, write the next three terms and the rule.

   a)   2, 4, 8, 16 ...   b)   1, 3, 9, 27 ...   c)   2, 10, 50 ...   d)   400, 200, 100 ...
   e)   3645, 1215, 405, 135 ... f)   70 000, 7000, 700 ...   g)   4, 20, 100 ...   h)   288, 144, 72 ...
   i)   7, 21, 63 ...   j)   3125, 625, 125 ...   k)   5120, 1280, 320 ...   l)   4, 16, 64 ...

3   A sequence has first term 19 and rule 'add 18 each time'. What is the fourth term in this sequence?

4   A sequence has first term 23 and rule 'subtract 4·5 each time'. What is the third term in this sequence?

5   The first and third terms of a particular sequence are 5 and 45 respectively. The rule for this sequence involves multiplying by a positive number. What is the second term in the sequence?

6   The fourth and fifth terms of a sequence that involves subtraction are 97 and 78. What is the first term of the sequence?

→

7   The Fibonacci sequence starts with two ones. After that, each term is the sum of the previous two terms. So the third term is 1 + 1 = 2, the fourth term is 1 + 2 = 3 and so on. The first five terms in the Fibonacci sequence are 1, 1, 2, 3, 5. Find the next five terms.

>> Square numbers

The sequence of square numbers can be formed by squaring the positive integers (1, 2, 3 …) to give 1, 4, 9 ….

Exercise 9B

1   Copy and complete this table to list the first ten square numbers.

n	1	2	3	4	5	6	7	8	9	10
nth square number	1	4	9							

2   Note the differences between consecutive square numbers. What pattern can you spot?

3   Draw a table to list the 11th to 20th square numbers; knowing these can be really helpful in mathematics.

4   Sharon says that the square of an odd number will always be odd. Is she correct? Why?

5   Look at your table and you will notice that 9 + 16 = 25 ($3^2 + 4^2 = 5^2$). So 3, 4 and 5 are a **Pythagorean triple**. Find five more Pythagorean triples in your tables and write them in the form $a^2 + b^2 = c^2$.

>> Triangular numbers

This pattern shows the first four triangular numbers.

Exercise 9C

1   Copy the diagram above and continue the pattern up to the sixth triangular number.

2   Copy and complete the table below to list the first ten triangular numbers.

n	1	2	3	4	5	6	7	8	9	10
nth triangular number	1	3	6							

3   What is the difference between:
   a)   the 3rd and 4th triangular numbers   b)   the 7th and 8th triangular numbers
   c)   the 9th and 10th triangular numbers   d)   the 53rd and 54th triangular numbers?

4   Add together pairs of consecutive triangular numbers and note the answers. What do you notice? Try to use your diagrams from question 1 to demonstrate why this result is true. Two colours of pen may be helpful!

5   Using what you observed in question 4, write down the sum of
   a)   the 11th and 12th triangular numbers   b)   the 19th and 20th triangular numbers.

6   James tries adding together the first square and triangular numbers, then the second and so on. He thinks that this will always result in a prime number. Is he correct? Give a reason for your answer.

▶ Linear patterns

In a **linear sequence** there is a **common difference** between consecutive terms.

Worked example   Decide whether each sequence is linear.

a) 6, 17, 28, 39 …

The difference between any two consecutive terms is 11 so the sequence **is** linear.

b) 8, 3, −2, −7 …

The difference between any two consecutive terms is −5 so the sequence **is** linear.

c) 6, 12, 24, 48 …

The difference between the 1st and 2nd terms is 6, but the difference between the 2nd and 3rd terms is 12 so the sequence **is not** linear.

Exercise 9D

1 Decide whether each sequence is linear and, if so, state the common difference.

   a) 17, 23, 29, 35 …   b) 2, 4, 8, 16 …   c) 18, 8, −2, −12 …   d) 1, 4, 9, 16 …   e) 7, 15, 23, 31 …

   f) 1, 8, 27 …   g) $\frac{1}{7}, \frac{3}{7}, \frac{5}{7}, 1$ …   h) $\frac{3}{5}, 1, 1\frac{2}{5}, 1\frac{4}{5}$ …   i) 3, 4, 6, 9 …   j) −19, −7, 5 …

It is more convenient to write down an algebraic rule to define a sequence, rather than listing terms. This rule is the **general term**. To calculate a particular term in a sequence, substitute the term number into the general term, taking care with the order of operations.

Worked example   A particular sequence has the general term $8n − 3$. Find:

a) the first term

For the first term, $n = 1$

$8n − 3 = 8 \times 1 − 3$

$= 8 − 3$

$= 5$

b) the fifth term

For the fifth term, $n = 5$

$8n − 3 = 8 \times 5 − 3$

$= 40 − 3$

$= 37$

c) the 100th term.

For the 100th term, $n = 100$

$8n − 3 = 8 \times 100 − 3$

$= 800 − 3$

$= 797$

---

2 Calculate the first term for each of these general terms.

   a) $4n + 1$ b) $6n − 8$ c) $3n + 5$ d) $9n + 11$ e) $7n − 4$ f) $5n − 9$

   g) $−9n + 6$ h) $−14n + 15$ i) $0{\cdot}5n + 8$ j) $−0{\cdot}2n + 12$ k) $\frac{2}{5}n − 3$ l) $\frac{3}{4}n + \frac{1}{2}$

3 For each general term in question 2, calculate the 5th term and the 20th term.

4 a) Find the first four terms of each of the sequences with these general terms.

      i) $4n + 5$ ii) $6n − 1$ iii) $−5n + 4$

   b) What connection can you spot between the common difference and the general term of each sequence?

To find the general term for a linear pattern, follow these steps.

- Find the common difference.
- Multiply the term number n by the common difference.
- Look for the adjustment required to match the **nth term**.
- Write down the general term

Worked example   Complete the tables and find a general term for each sequence.

a)

n	1	2	3	4	5	6
nth term	6	12	18	24	30	36

Common difference = 6
Multiples of 6: 6, 12, 18, 24 ...
No adjustment required
General term is $6n$.

b)

n	1	2	3	4	5	6
nth term	4	7	10	13	16	19

Common difference = 3
Multiples of 3: 3, 6, 9, 12 ...
Adjustment required: +1
General term is $3n + 1$

c)

n	1	2	3	4	5	6
nth term	5	12	19	26	33	40

Common difference = 7
Multiples of 7: 7, 14, 21, 28 ...
Adjustment required: −2
General term is $7n - 2$

d)

n	1	2	3	4	5	6
nth term	7	5	3	1	−1	−3

Common difference = −2
Multiples of −2: −2, −4, −6, −8 ...
Adjustment required: +9
General term is $-2n + 9$ (notice the sequence goes down when multiplying by a negative)

---

5 For each sequence, copy and complete the table and find the general term.

a)

n	1	2	3	4	5	6
nth term	7	14	21			

b)

n	1	2	3	4	5	6
nth term	3	6	9			

c)

n	1	2	3	4	5	6
nth term	12	24	36			

d)

n	1	2	3	4	5	6
nth term	−4	−8	−12			

e)

n	1	2	3	4	5	6
nth term	5	7	9			

f)

n	1	2	3	4	5	6
nth term	6	10	14			

g)

n	1	2	3	4	5	6
nth term	7	10	13			

h)

n	1	2	3	4	5	6
nth term	16	23	30			

i)

n	1	2	3	4	5	6
nth term	2	5	8			

j)

n	1	2	3	4	5	6
nth term	2	13	24			

k)

n	1	2	3	4	5	6
nth term	3	12	21			

l)

n	1	2	3	4	5	6
nth term	1	6	11			

m)

n	1	2	3	4	5	6
nth term	5	2	−1			

n)

n	1	2	3	4	5	6
nth term	11	7	3			

o)

n	1	2	3	4	5	6
nth term	−4	−7	−10			

p)

n	1	2	3	4	5	6
nth term	−13	−18	−23			

▶ Determining a formula

Worked example The first four matchstick patterns in a particular sequence are shown below.

a) Draw the next matchstick pattern in the sequence.

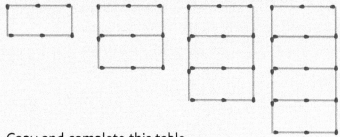

b) Copy and complete this table.

We do not need to write down terms 7 to 9.

Number of rectangles (r)	1	2	3	4	5	6	...	10
Number of matchsticks (m)	6	10	14	18	22	26	...	42

To find the number of matchsticks needed for the 10th pattern, note the common difference and count on from the 6th pattern.

c) Find a formula for the number of matchsticks m required for r rectangles.
Find the general term (in terms of r) for the sequence.
Write your formula to find the number of matchsticks using the letters given in the question.
$m = 4r + 2$

d) Use your rule to find the number of matchsticks needed for 14 rectangles.

$r = 14$
$m = 4 \times 14 + 2$
$\quad = 56 + 2$
$\quad = 58$
58 matchsticks are needed.

e) One particular pattern uses 126 matchsticks. How many rectangles are in this pattern?

$m = 126$
$126 = 4r + 2$
$\underline{-2 \qquad -2}$
$\quad 4r = 124$
$\underline{\div 4 \qquad \div 4}$
$\quad\quad r = 31$
There are 31 rectangles in the pattern.

Exercise 9E

1 Copy this matchstick pattern.

    a) Draw the next matchstick pattern in the sequence.

    b) Copy and complete this table.

Number of squares (s)	1	2	3	4	5	6	...	10
Number of matchsticks (m)	4	7	10				...	

    c) Find a formula for the number of matchsticks *m* required for *s* squares.

    d) Use your formula to find the number of matchsticks needed for 25 squares.

    e) One particular pattern uses 103 matchsticks. How many squares are in this pattern?

2 Copy this matchstick pattern.

    a) Draw the next matchstick pattern in the sequence.

    b) Copy and complete this table.

Number of triangles (*t*)	1	2	3	4	5	6	...	10
Number of matchsticks (*m*)	3	5	7				...	

    c) Find a formula for the number of matchsticks *m* required for *t* triangles.

    d) Use your formula to find the number of matchsticks needed for 18 triangles.

    d) One particular pattern uses 47 matchsticks. How many triangles are in this pattern?

3 Temperature is often measured either in degrees Celsius or in degrees Fahrenheit.

Degrees Celsius (*C*)	1	2	3	4	5
Degrees Fahrenheit (*F*)	33·8	35·6	37·4		

    a) Copy and complete the table.

    b) Use the table to find a formula connecting the two temperature scales.

    c) Convert 20 degrees Celsius to degrees Fahrenheit.

    d) Convert 50 degrees Fahrenheit to degrees Celsius.

    e) What is the freezing point of water in degrees Fahrenheit?

4 The relationship between the number of sides of a polygon and the sum of its interior angles form a linear pattern.

Number of sides (*s*)	3	4	5	6	7	8
Angle sum (*a*°)	180°	360°	540°			

    a) Copy and complete the table.

    b) Use the table to find a formula for calculating the angle sum of a polygon with *s* sides.

    c) What is the angle sum of a dodecagon (a polygon with 12 sides)?

    d) A given polygon has an angle sum of 3240°. How many sides does it have?

    e) What happens if you let *s* = 2 in your formula? Why?

5 Draw a set of axes with 0 to 6 along the *x*-axis and 0 to 24 along the *y*-axis. Look at the table you completed in question 1 b) and plot the number of squares along the *x*-axis and the number of matchsticks along the *y*-axis. For example, the first point will be (1, 4). Use a ruler to join up your points: what do you notice? Extend this line so that it reaches the *y*-axis. Can you spot any connections between your graph and the rule you wrote down? Try repeating this process for the table in question 2. What do you find?

Check-up 👍

1. For each sequence
   - i) write the next three terms in the sequence
   - ii) describe in words a rule for the sequence.
   - a) 4, 9, 14, 19 …
   - b) 28, 25, 22, 19 …
   - c) 3, 6, 12 …
   - d) 15, 3, −9 …
   - e) 128, 64, 32 …
   - f) 7, 10·7, 14·4, 18·1 …
   - g) $11, 10\frac{1}{4}, 9\frac{1}{2}, 8\frac{3}{4}$ …
   - h) 2, −8, 32 …

2. The 4th, 5th and 6th terms of a particular sequence are −9, −2 and 5. What is the first term of this sequence?

3. The 2nd, 4th and 6th terms of a particular sequence are 23, 31 and 39. What is the first term of this sequence?

4. What is the sum of the 3rd and 6th square numbers?

5. What is the sum of the first four triangular numbers?

6. Find two square numbers that add up to 80.

7. Find the first two triangular numbers that differ by 18.

8. Copy the grid. Cross out every square number, triangular number and Fibonacci number. What is the sum of the remaining numbers?

1	49	3	5	6
22	81	21	16	25
50	10	40	9	2
55	8	64	4	19
7	36	34	121	15

9. Calculate the first term for the sequence with the given general term.
   - a) $4n + 12$
   - b) $9n − 8$
   - c) $−10n − 4$
   - d) $−9n + 12$

10. For each of the general terms in question 9 calculate the third term and the one hundredth term.

11. The general term for the nth triangular number is $\frac{1}{2}n(n + 1)$. Use this formula to find
    - a) the 20th triangular number
    - b) the 24th triangular number
    - c) the 100th triangular number.

12. Copy and complete each table and find the general term for the sequence.

a)

n	1	2	3	4	5	6
nth term	20	27	34			

b)

n	1	2	3	4	5	6
nth term	5	12	19			

c)

n	1	2	3	4	5	6
nth term	46	96	146			

d)

n	1	2	3	4	5	6
nth term	7·5	8	8·5			

13. Copy and complete each table and find the general term for the sequence.

a)

n	1	2	3	4	5	6
nth term	15	11	7			

b)

n	1	2	3	4	5	6
nth term	6	4	2			

c)

n	1	2	3	4	5	6
nth term	−5	−7	−9			

d)

n	1	2	3	4	5	6
nth term	−16	−27	−38			

14 Look at this matchstick pattern.

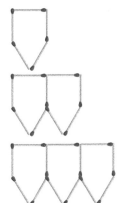

a) Draw the next matchstick pattern in the sequence.

b) Copy and complete this table.

Pattern number (*n*)	1	2	3	4	5	6	···	10
Number of matchsticks (*m*)	5	9	13				···	

c) Find a formula for the number of matchsticks *m* required for the *n*th pattern.

d) Use your formula to find the number of matchsticks needed for the 23rd pattern.

e) Which pattern in the sequence requires exactly 77 matchsticks?

15 Look at this matchstick pattern.

a) Draw the next matchstick pattern in the sequence.

b) Copy and complete the following table.

Pattern number (*n*)	1	2	3	4	5	6	···	10
Number of matchsticks (*m*)	7	12	17				···	

c) Find a formula for the number of matchsticks *m* required for the *n*th pattern.

d) Use your formula to find the number of matchsticks needed for the 19th pattern.

e) Which pattern in the sequence requires exactly 127 matchsticks?

16 A mobile phone contract with 1 GB of data costs £27. The same network offer a 2 GB contract for £35, a 3 GB contract for £43 and so on.

a) What is the additional cost for each additional gigabyte (GB) of data?

b) What is the fixed charge for a contract?

c) Find a formula to work out the cost (*c*) of a mobile phone contract given the amount of data (*d*) (in gigabytes).

d) Use your rule to work out the contract with the largest whole number of gigabytes that can be purchased for £100.

17 A company makes modular furniture that can be fitted together to suit the buyer's needs. Their most popular product is a shelving unit, which can be combined with other units of the same type. Adding more shelving units requires additional wooden boards and screws as shown in the table below.

Number of shelving units (*u*)	1	2	3	4	5
Number of wooden boards (*w*)	16	27	38		
Number of screws (*s*)	24	48	72		

a) Copy and complete the table.

b) Find separate formulae for the number of wooden boards and the number of screws needed given the number of shelving units required.

c) A customer buys material to combine 8 shelving units. They open the box to find 189 screws and 93 wooden boards. Is this enough to build their shelving units?

▶ The bridges of Königsberg

In the 18th century, the people of a town called Königsberg were puzzled. They wanted to know if they could visit each area of their town, crossing each bridge only once.

The plan below represents their town. It had four regions to visit and seven bridges linking them.

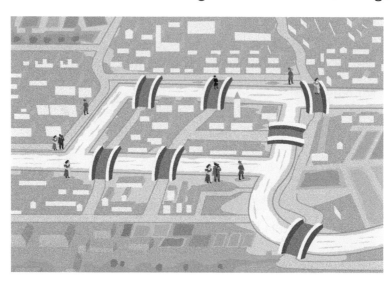

1   Can you find a route around the town that involves crossing every bridge only once?

Look at the diagrams below. They are called graphs, the dots are vertices and the connecting lines are edges. Graphs like these were first drawn in 1736 by Leonard Euler to analyse the bridges problem.

2   Copy each diagram. Try to draw it without lifting your pencil from the paper or tracing any edge more than once. Write 'Yes' or 'No' under each one. Think about what makes it work or stops you.

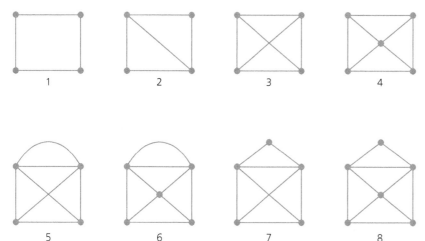

(**Hint**: Count the edges around each vertex. Is it an odd or even number?

When you get to a vertex, how do you get out? Do you need to come back?)

3   Keep investigating! Continue the pattern of shapes above or try the ones below. You can also make up your own.

Leonard Euler (and maybe you) realised that:

●   when edges come in even pairs then one can lead into a vertex and one can lead out, no problem!

●   only the start and end vertex can have an odd number of edges.

The last diagram in question 3 is just like the bridges of Königsberg.

All four vertices (regions) have an odd number of edges (bridges) so the journey is impossible!

4   Now that you know the rule, test it. Decide whether the graphs below are traceable by counting the edges around the vertices.

●   If there are 0 'odd' vertices then you should be able to trace it.

●   If there are exactly 2 odd vertices you should be able to trace it (and you know where to start and finish!).

●   Otherwise, it will be impossible.

Euler's elegant solution to the puzzle grew into a beautiful and complex branch of mathematics called graph theory. The diagrams display and explore the essential connections between things and have many applications in the world around us.

Euler's graphs can be used to model and understand the properties of many of the complex networks that are parts of our 21st-century life. Graph theory is used to analyse social media networks and the connections that the internet creates between millions of people. We live in homes that are connected to power networks and use computers

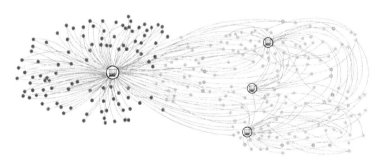

and electronics that are designed with efficient circuitry, using graph theory. We are lucky enough to have smartphones that use graph theory to find the fastest route between destinations at the touch of a button.

Graph theory has become a toolkit for studying the amazing networks inside our bodies, brains and DNA. It is astounding to think that all of this began with a question that puzzled the people strolling around Königsberg 300 years ago.

11 Coordinates, symmetry and scale

▶ Coordinates

We use **coordinates** to specify an object's position. You should already be familiar with reading coordinates on a grid and plotting points.

When writing down the coordinates of a point remember to:

- Use brackets
- Write the *x*-coordinate (the horizontal position) followed by the *y*-coordinate (the vertical position)
- Separate the *x*- and *y*-coordinates with a comma.

When drawing a coordinate grid it is important to:

- Use a ruler
- Label the axes
- Space out numbers equally along axes
- Write the number **on the lines**, not in the boxes. This allows us to be accurate.

Worked examples

1 Look at the grid and write down the coordinates of each point.

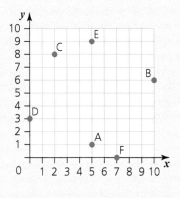

A is 5 along the *x*-axis and 1 up the *y*-axis so the coordinates of A are (5, 1). We write A(5, 1).

B is 10 along the *x*-axis and 6 up the *y*-axis, so B has coordinates (10, 6). We write B(10, 6).

In the same way, the remaining points have coordinates

C(2, 8)

D(0, 3)

E(5, 9)

F(7, 0).

2 Draw a coordinate grid, numbering the *x*- and *y*-axes from 0 to 10.

   a) Plot the points A(1, 7), B(1, 2) and C(4, 2).

   b) Plot the point D such that ABCD is a rectangle and write down its coordinates.

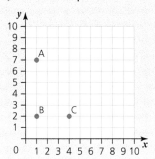

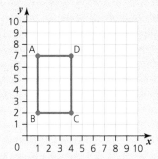

When plotting the points, be sure to read coordinates in the correct order. For example, A has coordinates (1, 7), so we go 1 along and 7 up and mark the point.

Once we have plotted A, B and C we can use a pencil and ruler to draw the rectangle containing those points and mark the final **vertex** (corner) D, which has coordinates D(4, 7).

Exercise 11A

1   Look at the grid.

   a) Write down the coordinates of each point.

   b) Which two points have the same x-coordinate?

   c) Which two pairs of points have the same y-coordinate?

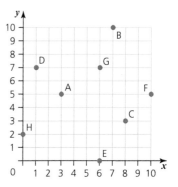

2   Draw a coordinate grid like the one in question **1** and plot these points.

   A(2, 5)   B(7, 1)   C(6, 8)   D(0, 4)   E(1, 1)   F(5, 0)   G(8, 3)   H(2, 10)

   I(4, 7)   J(2, 0)   K(7, 8)   L(0, 9)   M(10, 1)   N(1, 6)   O(0, 0)   P(3, 5)

3   Draw another coordinate grid of the same size.

   a) Plot the points A(1, 5) and B(10, 8) and join them with a straight line.

   b) Plot the points C(2, 9) and D(8, 0) and join them with a straight line.

   c) Write down the coordinates of the point where lines AB and CD cross over. We call this the
          point of intersection.

4   Plot each pair of points, connect them with straight lines and write down the coordinates of the
    point of intersection.

   a) E(0, 5) and F(10, 3)   **b)** I(1, 2) and J(9, 10)   **c)** M(1, 6) and N(9, 2)   **d)** Q(6, 10) and R(3, 1)

   G (1, 10) and H(7, 1)   K(5, 9) and L(5, 3)   O(8, 3) and P(2, 3)   S(9, 4) and T(1, 10)

5   If two points lie on the same vertical line, what can we say about their coordinates?

6   If two points lie on the same horizontal line what can we say about their coordinates?

7   Each of the coordinate axes shown has a mistake. Look at each one and describe the mistake that
    has been made.

a) 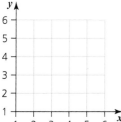   b) 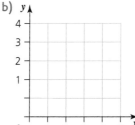   c) 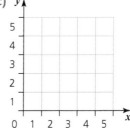   d)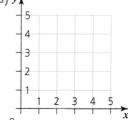

8   Draw a coordinate grid, numbering the x- and y-axes from 0 to 10.

   a) Plot the points A(3, 5), B(3, 2) and C(6, 2).

   b) Plot the point D such that ABCD is a square and write down its coordinates.

9 Draw a coordinate grid, numbering the x- and y-axes from 0 to 10.

   a) Plot the points A(1, 2), B(4, 5) and C(5, 2).

   b) Plot two possible locations for point D so that A, B, C and D are the vertices of a parallelogram.

10 Draw a coordinate grid, numbering the x- and y-axes from 0 to 10.

   a) Plot the points A(6, 2), B(1, 7) and C(3, 9).

   b) Plot the point D such that ABCD is a rectangle and write down its coordinates.

11 Draw a coordinate grid, numbering the x- and y-axes from 0 to 10.

   a) Plot the points A(3, 7), B(7, 7) and C(10, 3).

   b) Plot the point D such that ABCD is a trapezium (with a vertical line of symmetry) and write down its coordinates.

We can extend our coordinate grids to comprise four **quadrants** instead of one. To do so, continue the x- and y-axes beyond the origin to include negative numbers. The rules for writing and reading coordinates remain the same:

- Use brackets

- Separate coordinates by a comma

- Write the x-coordinate followed by the y-coordinate.

Worked example

Look at the grid and write down the coordinates of each point.

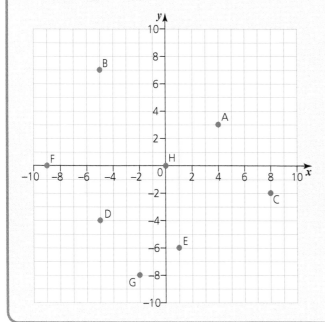

A is 4 in the **positive** x-direction and 3 in the positive y-direction, so A has coordinates (4, 3).

We write A(4, 3).

B is 5 in the **negative** x-direction and 7 in the positive y-direction so B has coordinates (−5, 7).

We write B(−5, 7).

C is 8 in the positive x-direction and 2 in the negative y-direction so we write C(8, −2).

D is 5 in the negative x-direction and 4 in the negative y-direction so we write D(−5, −4).

In the same way we have

E(1, −6), F(−9, 0), G(−2, −8) and H(0, 0).

Exercise 11B

1 Look at the grid.

a) Write down the coordinates of each point.

b) Which three points have a y-coordinate of −3?

c) Which point has prime numbers as both coordinates? (Remember: prime numbers must be positive!)

d) Besides the points in question b), there are two other sets of three points that lie on straight lines. What are they?

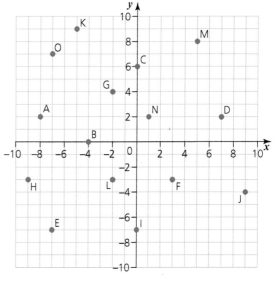

2 Draw a coordinate grid like the one in question 1 and plot these points.

A(−2, 6)   B(3, −4)   C(−7, −5)   D(0, −8)   E(7, 9)   F(−8, 3)   G(−10, −9)   H(7, −2)

I(−4, 0)   J(1, 6)   K(5, −7)   L(−9, −7)   M(4, −10)   N(10, 1)   O(−6, 6)   P(−3, 8)

3 Look at the coordinate grid and use it to decipher the messages by looking up the coordinates and writing down the corresponding letter.

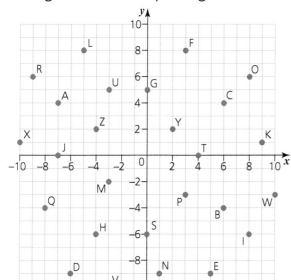

a) (4, 0), (−9, 6), (−7, 4), (3, −3), (5, −9), (−4, 2), (8, −6), (−3, 5), (−3, −2)

b) (8, −6), (1, −9), (4, 0), (5, −9), (0, 5), (5, −9), (−9, 6), (0, −6)

c) (−8, −4), (−3, 5), (−7, 4), (−6, −9), (−9, 6), (8, −6), (−5, 8), (−7, 4), (4, 0), (5, −9), (−9, 6), (−7, 4), (−5, 8)

Now interpret the coordinates and then rearrange the letters to get the name of a famous mathematician.

d) (2, 2), (−7, 4), (−7, 4), (4, 0), (−4, −6), (3, −3), (8, −6)

e) (6, −4), (−7, 4), (6, 4), (8, 6), (3, 8), (8, −6), (6, 4), (8, −6), (1, −9)

f) (−9, 6), (−7, 4), (1, −9), (8, −6), (−5, 8), (4, 0), (0, 5), (−7, 4), (−3, 5), (1, −9)

4 Use the grid in question 3 to encode these words.

a) Ruler   b) Multiple   c) Prime number   d) Symmetry

5 Use the grid in question 3 to encode some mathematical words for a classmate.

▶ Lines of symmetry

A shape has a line of **symmetry** if we can draw a straight line across it so that one side is a **reflection** of the other. If you were to fold a shape along a line of symmetry both sides would match up exactly. If a shape has symmetry this gives us information about the lengths of the sides and the sizes of the angles.

Worked example

For each shape, mark on all lines of symmetry and label any equal sides or angles.

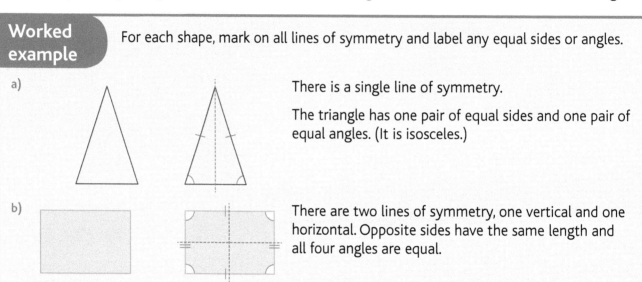

a) There is a single line of symmetry.

The triangle has one pair of equal sides and one pair of equal angles. (It is isosceles.)

b) There are two lines of symmetry, one vertical and one horizontal. Opposite sides have the same length and all four angles are equal.

Exercise 11C

1   Copy and name each shape. Draw on all the lines of symmetry and label any equal sides or angles.

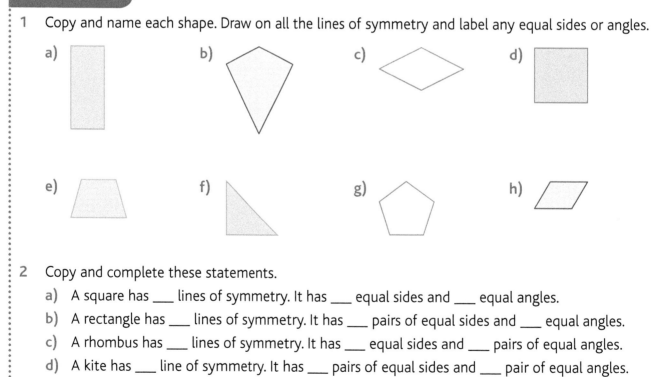

a)   b)   c)   d)

e)   f)   g)   h)

2   Copy and complete these statements.

a)   A square has ___ lines of symmetry. It has ___ equal sides and ___ equal angles.

b)   A rectangle has ___ lines of symmetry. It has ___ pairs of equal sides and ___ equal angles.

c)   A rhombus has ___ lines of symmetry. It has ___ equal sides and ___ pairs of equal angles.

d)   A kite has ___ line of symmetry. It has ___ pairs of equal sides and ___ pair of equal angles.

e)   A parallelogram has ___ lines of symmetry. It has ___ pairs of equal sides and ___ pairs of equal angles.

➔

3   For each picture, write the number of lines of symmetry.

a)   b)   c)   d)

e)   f)   g)   h)

4   In a **regular polygon**, all the sides are equal and all the angles are equal. Here are four regular polygons.

Equilateral triangle   Regular pentagon   Regular hexagon   Regular octagon

a)   Write down the number of lines of symmetry for each shape.

b)   How many lines of symmetry would you expect a regular heptagon (7 sides) to have?

c)   How many lines of symmetry would you expect a regular decagon (10 sides) to have?

Rangoli is an art form originating in India and involves colourful patterns, often with lines of symmetry. Here is an example of how to make a Rangoli-style pattern.

5   Draw a square grid, add its lines of symmetry, then draw a design in one portion of the square.

Reflect the design along a line of symmetry.

Continue to do this until the design is complete.

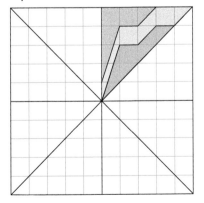
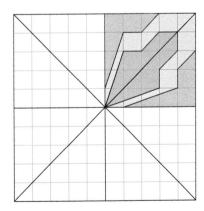

Try to create your own Rangoli-style pattern in the same way. Working through Exercise 11D will help.

▶ Creating symmetrical pictures

To complete a picture so that it has lines of symmetry:

- From the line of symmetry, 'count outwards' to each vertex of the picture.
- Take the same number of steps in the opposite direction from the line of symmetry and mark the **image** (reflection) of each vertex.
- Join the vertices with straight lines.

Worked example

Copy and complete each picture so that the dashed lines are lines of symmetry.

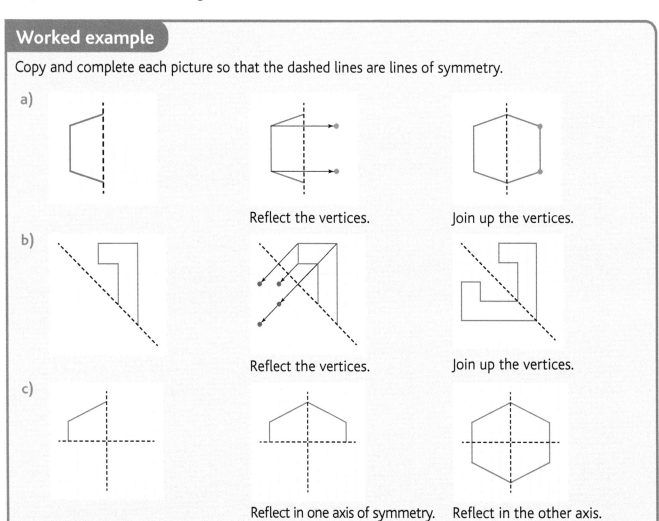

a)

Reflect the vertices.   Join up the vertices.

b)

Reflect the vertices.   Join up the vertices.

c)

Reflect in one axis of symmetry.   Reflect in the other axis.

Exercise 11D

1 Copy each diagram and complete it so that the vertical dashed line is a line of symmetry.

a)

b)

c)

d)

➜

e)   f)   g)   h)

2   Copy each diagram and complete it so that the horizontal dashed line is a line of symmetry.

a)   b)   c)   d)

e)   f)   g)   h)

3   Copy each diagram and complete it so that the diagonal dashed line is a line of symmetry.

a)   b)   c)   d)

4   Copy each diagram and complete it so that it is symmetric about the dashed lines.

a)   b)   c)   d)

e)   f)   g)   h)

i)   j)   k)   l)

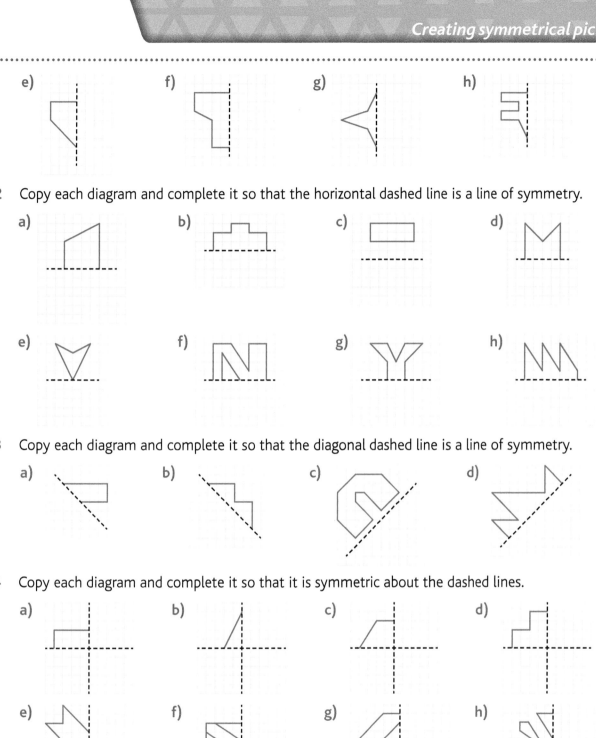

▶ Enlargement

When a shape is **enlarged** by a given amount, the lengths of its sides change but the size of each angle remains fixed. The degree of enlargement is known as the **scale factor**. To enlarge a shape by a given scale factor:

- Multiply the length of each side by the scale factor.
- Draw a copy of the shape but use the new lengths.

Given an object and an enlargement of the same shape, we can calculate the scale factor. To do so:

- Identify a pair of **corresponding lengths** (lengths in the same position).
- Divide the length in the large shape by the corresponding length in the smaller shape.

Worked examples

1   Enlarge this shape by a factor of 3.

The length of the triangle is 2 units and the height is 3 units. Multiplying by the scale factor:

$2 \times 3 = 6$

$3 \times 3 = 9$

So in our enlargement, the length should be 6 units and the height should be 9 units.

2   A square has sides of length 8 cm. An enlargement of the square has sides of length 40 cm. Calculate the scale factor of the enlargement.

$40 \div 8 = 5$. The square has been enlarged by a factor of 5.

Exercise 11E

1   Copy each shape and then enlarge them by the given scale factor.

a)

Scale factor: 2

b)

Scale factor: 3

c)

Scale factor: 2

d)

Scale factor: 4

e)

Scale factor: 3

f)

Scale factor: 2

g)

Scale factor: 2

h)

Scale factor: 3

→

2　A triangle has a base 10 cm and height 8 cm. An enlargement of the triangle has a base of 40 cm.

　　a)　What is the scale factor of the enlargement?

　　b)　What is the height of the enlarged triangle?

3　The diagonals of a rhombus are 16 cm and 9 cm long. An enlargement of the rhombus has its longest diagonal 48 cm.

　　a)　What is the scale factor of the enlargement?

　　b)　What length is the second diagonal in the enlarged rhombus?

4　A rectangle has a length of 18 cm and a breadth of 13 cm.
　　An enlargement of the rectangle has a breadth of 52 cm.

　　a)　What is the scale factor of the enlargement?

　　b)　What is the length of the enlarged rectangle?

　　c)　What is the area of the enlarged rectangle?

▶ Reduction

We can **reduce** a shape by a given scale factor in a similar fashion to enlargement. Reduction scale factors are usually written as fractions. To reduce a shape by a given scale factor:

- Multiply the length of each side by the given scale factor (find the fraction of each side).
- Draw a copy of the shape by using the reduced lengths.

To calculate a reduction scale factor:

- Identify a pair of corresponding lengths.
- Write the two lengths as a fraction with the shorter length as the numerator.
- Simplify if possible.

Worked examples

1　Reduce this shape by a factor of $\frac{1}{2}$.

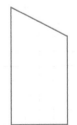

The length of the base is 4 units, the left side is 8 units and the right side is 6 units. Multiply each of these by the scale factor:

$4 \times \frac{1}{2} = 2$

$8 \times \frac{1}{2} = 4$

$6 \times \frac{1}{2} = 3$

So in our reduction, the base should be 2 units, the left side should be 4 units and the right side should be 3 units.

2　A triangle has a base of 18 cm and a reduction of the triangle has a base of 6 cm. Calculate the scale factor of the reduction.

$\frac{6}{18} = \frac{1}{3}$　The reduction scale factor is $\frac{1}{3}$.

Exercise 11F

1   Copy each shape and then reduce it by the given scale factor.

a)

Scale factor: $\dfrac{1}{2}$

b)

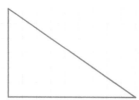

Scale factor: $\dfrac{1}{3}$

c)

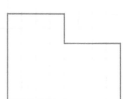

Scale factor: $\dfrac{1}{2}$

d)

Scale factor: $\dfrac{1}{4}$

e)

Scale factor: $\dfrac{1}{4}$

f)

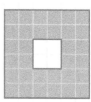

Scale factor: $\dfrac{1}{2}$

2   A rectangle has a length of 64 cm and a breadth of 28 cm. A reduction of the rectangle has a length of 16 cm.

   a)   Calculate the scale factor of the reduction.

   b)   What is the breadth of the smaller rectangle?

3   A triangle has a base of 49 cm and a height of 105 cm. A reduction of the triangle has a base of 7 cm.

   a)   Calculate the scale factor of the reduction.

   b)   What is the height of the smaller rectangle?

4   A trapezium has height 24 cm and base 60 cm. A reduction of the trapezium has a base of 40 cm.

   a)   Calculate the scale factor of the reduction.

   b)   What is the height of the smaller trapezium?

5   A triangle has height 25 cm and base 40 cm. A reduction of the triangle is drawn using a scale factor of $\dfrac{3}{5}$.

   a)   Calculate the height and base of the smaller triangle.

   b)   Calculate the area of the smaller triangle.

6   A square has sides of length 20 cm. A reduction of the square is drawn using a scale factor of $\dfrac{1}{5}$.

   a)   Calculate the length of the smaller square.

   b)   How many times larger is the area of the original square than that of the smaller one?

▶ Scale drawing

When an object is too large or small to be easily represented on paper we can use a scale drawing. This involves enlarging or reducing the original object by an appropriate scale factor and producing an accurate copy. Scale drawings are especially useful to architects, engineers and people producing maps.

We write a scale as a ratio, for example the scale of '1 cm : 100 m' means that every 1 cm in our scale drawing represents 100 m in reality.

Worked example

A school hall has a rectangular floor with dimensions as shown. Using a scale of 1 cm : 10 m, produce a scale drawing of the hall and use it to estimate the diagonal length of the real hall.

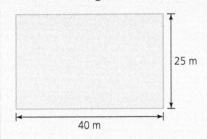

25 m

40 m

Use the scale to determine the dimensions of your scale drawing.

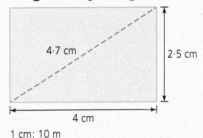

Our scale drawing should have a length of 4 cm and a breadth of 2·5 cm.

Draw your scale drawing carefully, being sure to show the scale.

4·7 cm   2·5 cm

4 cm

1 cm: 10 m

Measuring the diagonal of the scale drawing gives approximately 4·7 cm.

Using our scale, 4·7 × 10 = 47

The diagonal length of the real hall is approximately 47 m.

Exercise 11G

1   Make accurate scale drawings of these shapes. Use the scale shown under the shape.

a)

5 m

Scale: 1 cm : 1 m

b)

20 cm

40 cm

Scale: 1 cm : 5 cm

c)

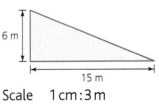

6 m

15 m

Scale   1 cm : 3 m

d)

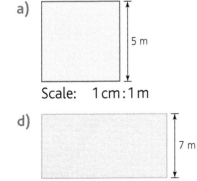

7 m

14 m

Scale: 1 cm : 2 m

e)

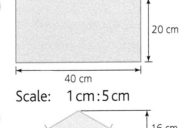

16 cm

56 cm

Scale: 1 cm : 8 cm

f)

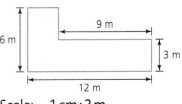

6 m   9 m

3 m

12 m

Scale: 1 cm : 2 m

2   In cricket, the wicket is a rectangular strip of the field approximately 3 m wide and 20 m long.

   a) Make a scale drawing of the wicket. Use a scale of 1 cm : 1 m.

The fastest recorded speed at which a cricket ball has been bowled is 161·3 km/h by Shoaib Akhtar in 2003.

   b) If a ball was bowled at this speed on the wicket described above, how long would it take to travel the length of the wicket? Give your answer in seconds. You may use a calculator. Remember: the batsman has slightly less than this amount of time to react!

3   This is a plan of a shopping centre car park.

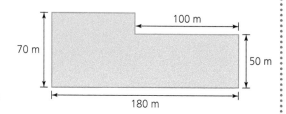

   a) Make a scale drawing of the car park. Use a scale of 1 cm : 10 m.

   b) Each parking space will need to be approximately 12 m². In theory, what is the maximum number of parking spaces that the carpark could hold?

   c) Give a reason why this number is likely to be an overestimate.

4   An NBA basketball court is a rectangle measuring 90 feet by 54 feet. An artist is hired to produce a design for the surface of the court. She decides to make a scale drawing, using a scale of 1 cm : 10 feet.

   a) What should the dimensions of her scale drawing be?

   b) Make an accurate scale drawing, using the scale given.

5   The maximum size of a regulation football pitch is a rectangle 120 m long and 90 m wide.

   a) Make a scale drawing of the football pitch, using a scale of 1 cm : 10 m.

   b) Measure the diagonal of your scale drawing.

   c) Use this to calculate the length of the diagonal of the real football pitch.

6   A hang glider is designed in the shape of an isosceles triangle with base 10·2 m and height 3·8 m. Make a scale drawing of the hang glider, using a scale of 1 cm : 500 cm. Be careful with units!

7   An architect's plan of a school hall is a rectangle measuring 24 cm long by 15 cm wide. The real length of the school hall is to be 48 m.

   a) Write down the scale of the drawing.

   b) Calculate the real width of the school hall.

8   A model Spitfire is 27 cm long and has a wingspan of 33 cm. A real Spitfire has a wingspan of 11 m.

   a) Write down the scale of the model.

   b) Calculate the real length of the Spitfire.

9   A toy brontosaurus is 44 cm long from head to tail. Brontosaurus fossils suggest they were 22 m from head to tail.

   a) Write down the scale of the toy as a ratio.

   b) A toy tyrannosaurus is made to the same scale and measures 24·6 cm long. Calculate the length of the real tyrannosaurus. Give your answer in metres.

Check-up

1   Look at the coordinate grid and write down the coordinates of each labelled point.

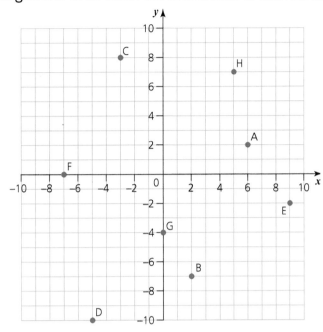

2   Draw a coordinate grid with axes numbered from 0 to 10.

   a)   Plot the points A(0, 4), B(5, 9) and C(8, 6).

   b)   ABCD is a rectangle. Plot the point D and write down its coordinates.

3   Draw a coordinate grid with axes numbered from −10 to 10 and plot these points.

   A(4, 8)   B(7, 9)   C(−2, 5)   D(3, −4)   E(0, −9)   F(−6, −1)
   G(5, −7)   H(−1, 0)   I(−6, 6)   J(10, 4)   K(8, −6)   L(−7, 3)

4   Draw a coordinate grid, numbering the x- and y-axes from 0 to 10.

   a)   Plot the points A(8, 4), B(3, 9) and C(1, 7).

   b)   Plot the point D such that ABCD is a rectangle and write down its coordinates.

5   Draw a coordinate grid, numbering the x- and y-axes from 0 to 10.

   a)   Plot the points A(2, 6), B(6, 9) and C(6, 4).

   b)   Plot the point D such that ABCD is a rhombus and write down its coordinates.

   c)   Draw in the diagonals of rhombus ABCD. What is their point of intersection?

6   Copy these shapes and complete them so that the dashed lines are lines of symmetry.

a)

b)

c)

d)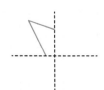

e)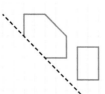

f)

7   How many lines of symmetry do these shapes have?

   a) rhombus   b) parallelogram   c) kite   d) equilateral triangle

8   Copy these shapes and enlarge them by the given scale factor.

   a)

   Scale factor: 3

   b)

   Scale factor: 2

9   Copy these shapes and reduce them by the given scale factor.

   a)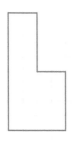

   Scale factor: $\dfrac{1}{2}$

   b)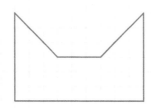

   Scale factor: $\dfrac{1}{3}$

10 A tennis court is a rectangle 78 feet long and 36 feet wide.

   a)   Using a scale of 1 cm : 6 feet, make a scale drawing of the tennis court.

   b)   Measure the diagonal of your scale drawing.

   c)   Use your answer to part b) to estimate the diagonal length of the real tennis court.

11 The sail of a boat is in the shape of a right-angled triangle, as shown.
   Make a scale drawing of the sail. Use a scale of 1 cm : 500 cm.

9 m

2·5 m

12 A souvenir model kit of the Eiffel Tower is 16·4 cm wide at its base. The real Eiffel Tower is 328 feet wide at its base.

   a)   Write down the scale of the model.

   b)   The model is 53 cm tall. Use your scale to calculate the real height of the Eiffel Tower.

12 Data analysis

▶ Categorical and numerical data

Statistics is one of the most widely used branches of mathematics. Everyone needs to understand the world around them and its resources. For example, athletes, climate scientists, engineers, city planners, medical researchers, retailers and politicians all use statistics.

The data we collect is either **numerical** or **categorical**.

- Numerical data takes a number value, e.g. height, weight, heart rate, temperature.
- Categorical data is a description of something, e.g. school subject, cloud formation, job, eye colour.

Exercise 12A

1 Make a table with the headings shown. List each data set under the correct heading.

Categorical	Numerical

time, calories, type of pet, share price, age, length, city, test score, wages, sport, career, savings, blood pressure.

2 Describe how you would investigate these questions. What kind of data would you collect?
   a) What sports do people in your year group enjoy?
   b) Which part of the UK has the highest mountains?
   c) What careers are S2 pupils interested in?
   d) Who has the fastest reaction times in your class?

▶ Compound bar graphs

A bar graph displays the frequency of categorical data using bars of different heights. A compound bar graph tells us more about the data in each category by breaking it down into groups. A key is included to identify the groups.

- Choose an appropriate and consistent scale.
- Use a ruler and plot data points accurately.
- Use colour and a key to identify groups.

The example shows a **clustered** compound bar graph. Each category has a cluster of bars.

Worked example

Draw a compound bar graph to display the information about molecules, which is shown in the table.

	Molecular formula	Number of carbon atoms	Number of hydrogen atoms	Number of oxygen atoms
Aspirin	$C_9H_8O_4$	9	8	4
Glucose	$C_6H_{12}O_6$	6	12	6
Starch	$C_6H_{10}O_5$	6	10	5

Show the molecules on the x-axis (horizontal). Each molecule will have a cluster of 3 bars.

Show the number of atoms on the y-axis (vertical). The maximum value is 12, go up in 1s.

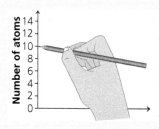

Label the tick marks on the grid line.

We are numbering every second box to make it easier to read.

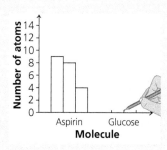

Be consistent with the width of the bars.

Leave a gap between clusters.

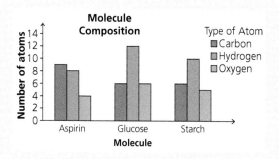

What story does the graph tell?

We can compare across the groups.

We can see the composition of glucose and starch is similar. They both have hydrogen as the most popular element.

The most common element of aspirin is carbon.

Exercise 12B

1   A supermarket keeps data on its employees at two branches.

Label the x-axis 'Branch' and the y-axis 'Number of employees'. Draw two bars (Ft and Pt) for the city branch and two for the local. Remember to include a key.

	Full-time	Part-time
City	8	14
Local	4	7

2   The table shows estimates for the sizes of the armies at the Battle of Bannockburn in 1314.

The data is based on the work of historians.

	Infantry	Cavalry	Bowmen
Scottish	5000	500	250
English	8000	2000	3000

Display the data in a clustered compound bar graph.

Label the x-axis 'Scottish and English Forces'. Draw two bars (Scottish and English) for each of the three kinds of force.

Label the y-axis 'Size of Force'. Go up in 500s. Remember to add a title, axes labels, colour and a key.

3 The pupils in a secondary school are divided into three house groups.

In Charity week S1, 2 and 3 raise money by doing sponsored events. The table shows the total raised by each class.

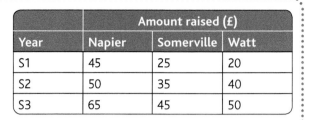

	Amount raised (£)		
Year	Napier	Somerville	Watt
S1	45	25	20
S2	50	35	40
S3	65	45	50

a) Display the results in a clustered compound bar graph.

Label the *x*-axis 'Year Group'.
Each year group will have a cluster of three bars.

Label the *y*-axis 'Amount Raised'.
The maximum is 65 so go up in 5s.

b) Which year group raised the most money?

c) Which house group raised the most money?

4 This graph shows information about cars for sale in a local dealership.

a) Which size of car is most popular with buyers?

b) Which fuel type is most popular with buyers?

c) How many SUVs are for sale in total?

d) What fraction of the SUVs are electric?

e) What fraction of the hatchbacks are electric?

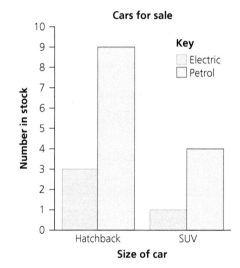

5 The graph shows information about ticket sales in a theatre.

a) How many tickets were sold for the upper circle on Friday?

b) Which performance sold the most tickets for the stalls?

c) What fraction of the audience at the Saturday matinee sat in the dress circle?

d) Which performance sold the most tickets in total?

e) Which part of the theatre sold the most tickets in total?

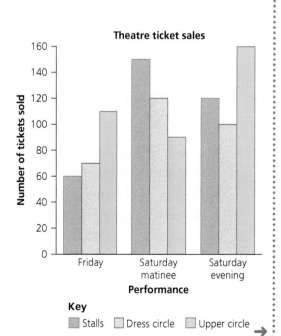

159

Question **6** shows a **stacked** compound bar graph.

6 An online business sells experience vouchers.

The graph shows information about the total value of their voucher sales for two months.

a) What was the total value of sporting activities vouchers sold in December?

b) What percentage of December's sales were for food and drinks vouchers?

c) How did the voucher sales change from December to January. Describe any changes you notice.

d) What fraction of January's voucher sales were for food and drink vouchers?

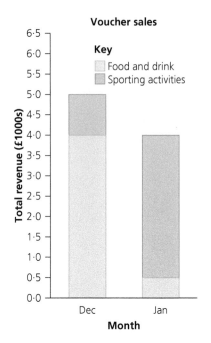

7 Priyanka and Lucy count the number of vegetarian options in two local restaurants.

a) Create a stacked compound bar graph to display the data.

	Vegetarian	Non-vegetarian
Italian	4	6
Japanese	10	5

b) Lucy is vegetarian and likes both cuisines. Which restaurant would you recommend for her?

c) What percentage of the menu options are vegetarian in each restaurant?

8 Two friends made a horizontal stacked bar chart to display the number of countries each had been to during their life.

a) What fraction of all of the countries Lizzie visited were in Europe?

b) There is no green on Carmel's bar. What does this mean?

c) Describe the differences between Carmel and Lizzie's travels.

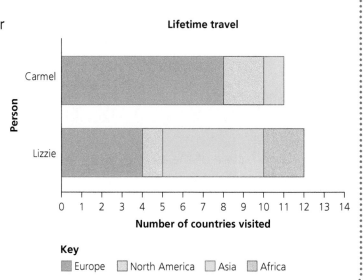

▶ Compound line graphs

Line graphs display change over a continuous numerical scale e.g. time, weight. Compound line graphs display results for more than one group. A key is included to identify the groups.

- Choose consistent scales that make good use of the space you have.
- Use colour and a key to identify groups.
- Plot data points accurately and use a ruler to join them up.

Worked example   A gardener tests a fertiliser called Bioforce. She plants two tomato plant seeds and feeds one with Bioforce.

Height (cm)	Age (days)	10	20	30	40	50	60
	Plant A (Bioforce)	10	30	70	85	88	90
	Plant B	5	10	20	50	80	100

She measures her tomato plants every 10 days for 60 days. The results are shown.

a) Display the results on a compound line graph.

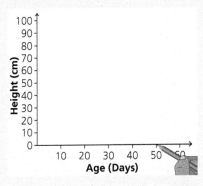 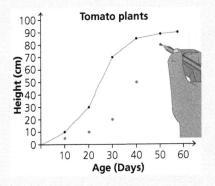

Age on the x-axis. Height on the y-axis. Height is the result of the experiment. Choose scales to use the space well.

Plot each point for plant A (10,10), (20, 30), ... and join these up with straight lines.

Do the same for plant B (10, 5), (20,10), ... join these with a different colour.

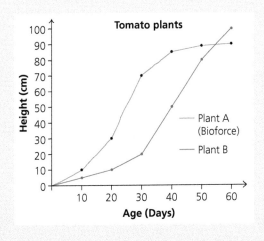

b) How tall was Plant A after 15 days?

Plant A was 20 cm tall.

c) When was the difference in heights greatest?

70 – 20 = 50 cm

At 30 days the gap between lines is largest, 50 cm.

d) After planting, when were the two plants the same height?

The lines cross at 55 days old, both almost 90 cm.

e) Do you think Bioforce works?

Plant A grew more quickly with Bioforce but Plant B was taller overall.

Exercise 12C

1 Andres is growing cress in different light conditions. He measures the height of the cress every 2 days for 14 days.

	Age (days)	0	2	4	6	8	10	12	14
Height (cm)	A Shaded	0	1	3	6	7	7	10	12
	B Sunlit	0	2	3	5	8	12	14	16

a) Follow the steps to draw a compound line graph to display the data.

Label the x-axis 'Age (days)'. The maximum is 14 days. Go up in 1s.

Label the y-axis 'Height (cm)'. The maximum is 16 cm. Go up in 1s.

Plot each point in turn for cress A. The first is (0, 0) then (2, 1), then (4, 3), (6, 6), ... Join these up with straight lines.

Plot each point for cress B. The first is (0, 0) then (2, 2) then (4, 3), ... Join these up with straight lines in a different colour.

Add a title and key to show which line is a) shaded and b) sunlit.

b) How tall was cress B on day 9?

c) Which cress was tallest on day 6?

d) On which two days after planting were plants A and B the same height?

e) What happened to cress A between days 8 and 10?

f) What did Andres learn about growing cress?

2 An S2 class is working on endothermic reactions. They mix solutions and measure the temperature using a digital thermometer probe. The table shows the results for two solutions.

	Time (mins)	0	1	2	3	4	5
Temperature °C	Solution 1	20	15	4	2	1	−1
	Solution 2	20	10	3	2	−1	−2

a) Draw a compound line graph to display the data.

The temperature falls below 0 °C. Set your axis up as shown.

Include a title, axis labels, colour and a key.

b) At which two times were the solutions the same temperature?

c) Which solution cooled to 0 °C first? Estimate when this happened.

d) What do you think an endothermic reaction is?

e) Which solution had the greatest drop in temperature in 1 minute? When was it?

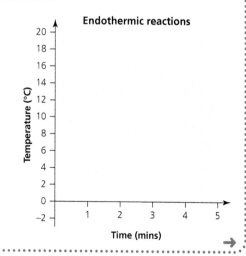

3 Look at the line graph below showing share prices.

a) What was the price of Scot Core at 14:10?

b) At what time was the price of Scot Core and Kendal Inc the same?

c) At what time was the difference in prices greatest?

d) The share price of Scot Core was fluctuating during this time. What does this mean?

e) Describe the trend in Kendal Inc share price. What happened at 14:15?

f) What would 500 shares in Kendal Inc cost at 14:00?

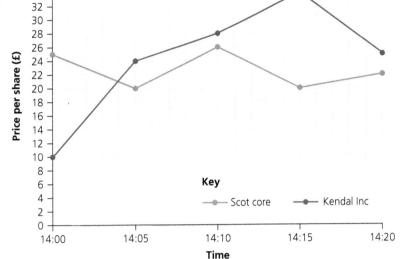

g) How much more would shares in Kendal Inc be worth at 14:20?

4 This compound line graph shows how the number of infants vaccinated against measles and hepatitis has changed over time. The data is sourced from the World Health Organization. The WHO uses data to work with governments around the world to promote good health and fight disease.

a) Estimate the percentage increase in measles vaccinations between 1980 and 1990.

b) What happened to the rate of measles vaccinations in 2000?

c) When was the hepatitis vaccine first approved?

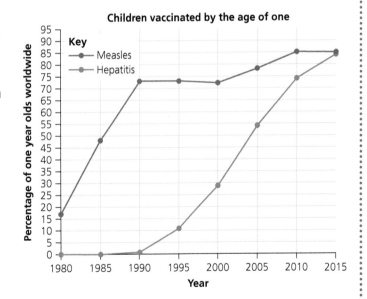

d) Estimate how far apart the rates of measles and hepatitis vaccinations were in 2005.

e) Which 5-year period saw the largest rise in the uptake of the hepatitis vaccine?

f) What happened in 2015?

▶ Reading pie charts

Pie charts show how a total is shared. The size of each slice represents how many data points are in each category (the frequency). To read a pie chart:

- Determine the size of the sector (slice), $\dfrac{x}{360}$, where $x°$ is the angle at the centre.
- Simplify this fraction if possible.
- Find that fraction of the data. This is what the slice represents.

Worked example A graduate trainee makes a pie chart to display her monthly budget of £1200.

a) How much is spent on bills?

The 'bills' sector has a right-angle at the centre, 90°

$$\frac{90}{360} = \frac{1}{4} \qquad 4\overline{)1200} \;\; 300$$

£300 on bills

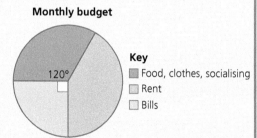

Monthly budget

Key
- Food, clothes, socialising
- Rent
- Bills

120°

b) How much does she spend on food, clothes and socialising?

This sector has 120° at the centre.

$$\frac{120}{360} = \frac{1}{3} \qquad 3\overline{)1200} \;\; 400$$

She spends £400 on food, clothes and socialising.

c) How much is her rent?

With the final slice deduct the total so far from the total budget.

$$\begin{array}{r} 400 \\ +300 \\ \hline 700 \end{array} \qquad \begin{array}{r} 1200 \\ -700 \\ \hline 500 \end{array}$$

£500 on rent

Exercise 12D

1 The pie chart shows the results of a survey of 200 pupils.

a) How many pupils walked to school?

b) How many pupils took the bus?

c) Show that $\dfrac{72}{360} = \dfrac{1}{5}$. How many pupils came by car?

d) How many pupils came by bike?

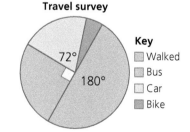

Travel survey

72°

180°

Key
- Walked
- Bus
- Car
- Bike

2 The pie chart shows data collected by a chain of salons about 600 customer visits.

a) How many people had a trim?

b) How many people had their hair coloured?

c) How many people had a restyle?

d) How many people had extensions?

Salon visits

60°

72° 120°

Key
- Trim
- Colour
- Restyle
- Extensions

3 A group of 900 school leavers were asked which mathematics careers they were interested in.

   a) How many people were interested in accountancy?

   b) How many people were interested in artificial intelligence?

   c) How many were interested in encryption?

   d) The 'strategist' slice represents 45 people. What is the angle at the centre of this slice?

   e) How many people were interested in being a data analyst?

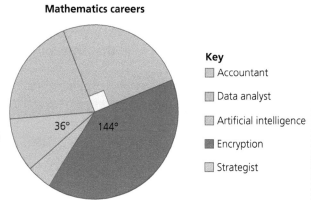

Mathematics careers

Key
- Accountant
- Data analyst
- Artificial intelligence
- Encryption
- Strategist

4 The pie chart shows the results of a survey about which sports people like to watch. The 'hockey' slice represents 80 people.

   a) How many people took the survey altogether?

   b) How many people voted for football?

   c) How many people voted for athletics?

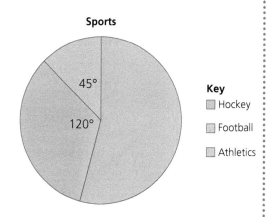

Sports

Key
- Hockey
- Football
- Athletics

If a pie chart displays percentages, we do not need the angle at the centre. Just calculate each percentage of the total.

5 The pie chart shows the destinations of Scottish exports in 2017. The data is from official Scottish government statistics. The total value of exports in 2017 was £80 billion to the nearest 10 million pounds. One billion is one thousand million.

   a) Find the total value of exports to Europe.

   b) Find the total value of exports to the rest of the world.

   c) Find the value of exports to the rest of the UK.

   d) What products do you think Scotland exports? Make a list.

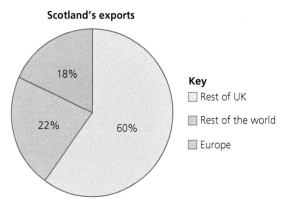

Scotland's exports

Key
- Rest of UK
- Rest of the world
- Europe

▶ Drawing pie charts

To draw a pie chart:
- Determine the size of the sector (slice) using **relative frequency** $= \dfrac{\text{frequency}}{\text{total}}$.
- Simplify the fraction if possible.
- Find that fraction of 360°. This is the angle you need to draw.

Worked example

Siobhan asks her friends what their favourite topic in mathematics is. Draw a pie chart to display her results.

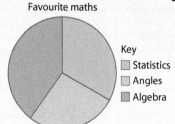

Favourite maths

Key
- Statistics
- Angles
- Algebra

	Friends (frequency)	Relative frequency	Angle at centre
Statistics	16	$\dfrac{16}{48} = \dfrac{1}{3}$	$\dfrac{1}{3}$ of 360° = 120°
Angles	12	$\dfrac{12}{48} = \dfrac{1}{4}$	$\dfrac{1}{4}$ of 360° = 90°
Algebra	20	$\dfrac{20}{48} = \dfrac{5}{12}$	$\dfrac{5}{12}$ of 360° = 150°
Total	48		

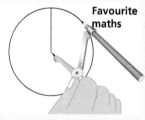

Favourite maths

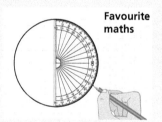

Favourite maths

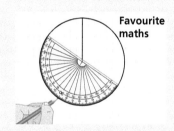

Favourite maths

Draw a circle, mark the centre and start the first sector.

Set the centre of your protractor over the centre of the circle. Measure the angle.

Keep the centre in place. Turn your protractor to measure 0° from the edge of the previous sector.

Exercise 12E

1 An athlete does 3 different kinds of training and records how many hours of each she does in a month.

a) Copy and complete the table.

b) Draw a pie chart to display the data.

	Hours (frequency)	Relative frequency	Angle at centre
Cardio	12	$\dfrac{\quad}{24} = \dfrac{\quad}{\quad}$	— of 360° = °
Strength	8	$\dfrac{\quad}{24} = \dfrac{\quad}{\quad}$	— of 360° = °
Dynamic	4	$\dfrac{\quad}{24} = \dfrac{\quad}{\quad}$	— of 360° = °
Total	24		

→

2 A school charity raises money from different events over a year.

	Amount raised (£s) (frequency)	Relative frequency	Angle at centre
Bag packing	400		
Talent show	360		
Fun run	240		
Smoothie Bike	200		
Total			

a) Copy and complete the table.

b) Draw a pie chart to display the data.

3 A class votes for their pupil council representative.

a) Copy and complete the table of results. Use a calculator to work out the angle.

	Votes (frequency)	Relative frequency	Angle at centre
Param	15		
Jodie	7		
Hector	11		
Total			

b) Draw a pie chart to display the data.

If the data is given as percentages we already have the relative frequency. To find the angle at the centre, just calculate each percentage of 360°.

4 The table shows how electricity was generated in Scotland in 2016.

Nuclear	43%
Coal	5%
Gas and oil	8%
Hydroelectric	12%
Other renewable	32%

The data is sourced from official Scottish Government Statistics. Create a pie chart to display the data.

You may use a calculator to find each percentage of 360°, round each to the nearest degree.

▶ Scatter graphs and correlation

Scatter graphs display possible links between two sets of data. For example: exam scores and study time, advertising budget and sales, age and strength. Scatter graphs display numerical data on both axes.

To make and read a scatter plot:

- Carefully draw and label the axes.
- Plot each data pair as a coordinate point.
- Look at the scatter graph. Is there a correlation? This means a relationship or trend, as shown in the example.
- Are there any outliers? These are points which don't follow the trend.

Worked examples

1 Comment on the relationship shown on each graph.

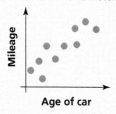

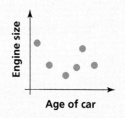

This is **positive correlation**.

Older cars have done more miles.

This is **negative correlation**. Older cars are worth less. There is an outlier. This is an old car but is worth a lot. Maybe it is vintage.

There is **no correlation** between a car's age and its engine size.

2 Some friends are keen to know how to do better in their mathematics tests. They carry out a study. Each friend chooses how long they will spend on their daily studying before the test. They agree to record the results shown. They each do their chosen amount of studying every day for 5 days before the test.

Time (mins)	5	15	7	10	30	25	18	4	20	22	30
Test Score (%)	50	80	40	65	90	100	70	80	90	75	100

a) Draw a scatter graph to display the data.

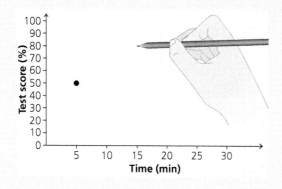

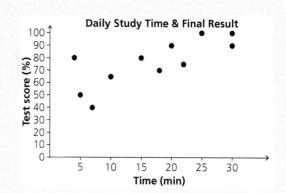

Time on the x-axis. Test score on the y-axis. The test scores are the result of the experiment.

Plot each point in turn (5, 50), (15, 80), (7, 40) but do not join up the dots.

b) Comment on the relationship between the variables.

There is a positive correlation. The pupils who spent longer on their daily study did better in the test.

There is an outlier who studied only 4 minutes and got 80%. Maybe they studied hard as they went along or maybe they weren't reporting all of their study time.

Exercise 12F

1 Each scatter graph displays 'time spent at work' on the x-axis. The y-axis labels are missing.

Choose a label for each y-axis from this list: 'time with family'   'wages'   'height'

a)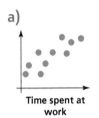

Time spent at work

b)

Time spent at work

c)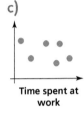

Time spent at work

2 Farah gathers some data about 10 of her classmates. The results are shown in the table.

Hours of exercise per week	2	1	3·5	2	4	2·5	3	5	4·5	3
Beep test score	7	8	9	8	10	11	9	12	11	8

a) Make a scatter graph to display the results. Display hours of exercise per week on the x-axis and beep test score on the y-axis. Plot each point in turn. The first is (2, 7) then (1, 8) etc.

b) Comment on the relationship between these variables. Remember to describe the type of correlation.

3 A café records data about the outside temperature and the number of salads they sell for one week.

Temperature (°C)	12	14	19	21	15	19	5
Number of salads sold	8	10	15	20	12	18	16

a) Make a scatter graph to display the data. Label the x-axis Temperature (°C) and the y-axis Number of salads sold. Plot each point in turn (12, 8), (14, 10) etc.

b) Comment on the relationship between these variables. Identify any outliers.

4 A charity has volunteers with collecting tins in a town centre. The charity records data about the total amount collected each month and the number of rainy days.

	Jan	Feb	March	April	May	June	July	August	Sept	Oct	Nov	Dec
Number of rainy days	19	15	12	18	13	9	10	23	25	21	22	25
Total donations (£s)	150	500	600	750	650	400	800	250	150	300	250	950

a) Draw a scatter graph to display the data.

b) Comment on the relationship shown. Identify any outliers.

c) Can we be sure that the rain is causing people to be less generous?

▶ Misleading statistics

Data is often used to inform opinions and make important decisions. We should always check the source of data.

Reliable sources do their best to gather data which fairly and accurately represents the world around us. They take care to gather enough data to support their conclusions and explain their methods so that everyone can judge for themselves. The Scottish Government and the World Health Organization are good examples of reliable sources.

When judging data, ask:

● What is the source of the data? Who gathered it and why?

● What is the sample size? How was the sample chosen? Does it represent the population?

● Look carefully at charts and graphs. How appropriate is the scale?

● Is this the whole picture? What other information do I need?

Exercise 12G

1 This is an advert for deodorant. How reliable do you think this data is?

*100 free samples issued, we advised customers to email us if they didn't like it.

> 9 out of 10 of you love* Freshies!

2 Joel is studying to be a vet. He is researching the health of local dogs by asking their owners questions about their wellbeing. Joel stands outside a local Puppy Parlour and interviews 7 dog owners.

 a) There are about 500 dogs in the local area. Do you think Joel has a large enough sample?

 b) Do you think choosing dogs from outside the Puppy Parlour is a good idea? Would these dogs represent the whole population fairly?

3 A deputy headmaster surveys pupils asking how happy they are with the food in the school canteen. He stands at the front of the queue and asks the first thirty pupils served how happy they are.

 a) Why might this survey be biased?

 b) Give two reasons that the deputy's sample will not represent the whole pupil population.

 c) Describe a better way to choose a random sample of pupils.

4 The pie chart shows the results of a pupil satisfaction survey.

 Give two reasons why this pie chart is misleading.

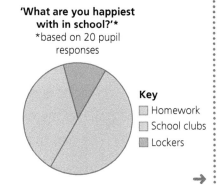

'What are you happiest with in school?'*
*based on 20 pupil responses

Key
☐ Homework
☐ School clubs
☐ Lockers

➡

5   The grouped bar chart claims to show that boys' attainment is well ahead of the girls.

Explain two things which are misleading about the *y*-axis.

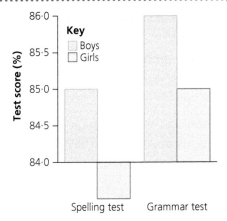

6   A website claims to have found out that people with bigger feet have slower reaction times. This graph is shown.

Does this graph support the claim? Can a scatter graph ever prove claims like these?

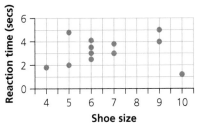

7   The head of a mathematics department asked 10 pupils, out of a year group of 300, what their favourite subject was. All of the 10 pupils did well in their exam. The results are shown.

State three problems with this study.

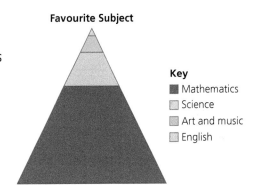

8   A manager keeps a record of the number of complaints they have received about their service. They display the results on the graph shown.

Why might this graph be misleading?

9   Create a misleading chart of your own. Carry out a small, biased survey or just make up any old rubbish! Present your 'findings' really nicely by producing a professional looking graph, chart or diagram.

Add a shocking title and build a class display of Fake News.

▶ Using technology

You have probably worked through most of this book with very little help from technology. Well done, your brain is amazing. When you learn new things and practise your skills by hand your brain grows and develops.

Technology can be a great tool for mathematicians. Computers can carry out millions of calculations in fractions of a second. They can also process large amounts of data quickly and create excellent charts and diagrams for us.

To use technology to produce charts and diagrams:

- Start with a table of data.

- Decide what kind of chart or diagram will be suitable.

- Is the data numerical or categorical?

- In Excel, highlight the data then choose > Insert > Chart > browse the chart options.

- Click to add a title, labels, change the scale, colours, etc.

- You can also look for an app or online chart maker, there are many which are free to use.

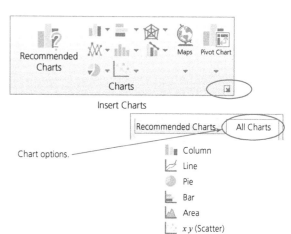

Exercise 12H

If you have access to technology, you can produce your own graphs for each question.

1   A secondary school has 4 houses and collects house points for achievements throughout the year. The results are shown in the spreadsheet.

a)   Which of the chart options above would you choose?

b)   Jamie tries to make a scatter graph. Explain why this will not work well.

	A	B
1	House	Points
2	Lomond	1241
3	Ness	1685
3	Morar	1332
4	Tay	1052

2   The data in the spreadsheet shows typical monthly temperatures in Glasgow and Mumbai.

	A	B	C	D	E	F	G	H	I	J	K	L	M
1		Jan	Feb	March	April	May	June	July	Aug	Sept	Oct	Nov	Dec
2	Glasgow °C	7	8	9	12	15	17	19	19	16	13	10	7
3	Mumbai °C	31	32	33	33	34	32	31	30	31	34	33	30

a)   Cara makes the line graph shown. What do you think of this line graph? What could Cara do better?

b)   Ash chooses to make a pie chart with temperature data. Explain why this will not work well.

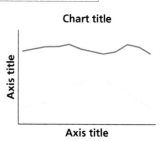

3   William is researching shipbuilding, one of Scotland's most important industries in the 19th and 20th Centuries. The table shows the number of ships built in three important shipbuilding centres during the decades shown. The data was gathered by historian Anthony Browning.

	A	B	C	D
1		Govan	Port Glasgow	Greenock
2	1840	65	40	136
3	1900	347	445	236
4	1960	81	99	60

Port Glasgow

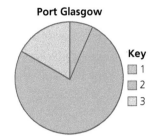

Key
□ 1
□ 2
□ 3

a)   William makes a pie chart using only the data for Port Glasgow. What could William do better?

b)   Can you think of a way to display all the data on one chart? What chart option would you choose? Use technology to produce a better display of the data or draw one by hand.

4   A psychologist tests the memory of volunteers at a science fair. They show the volunteers a box of 20 items and give them one minute to memorise them. Immediately after viewing, the volunteers must try to recall all the items. The number recalled and the age of the volunteers is recorded.

a)   What type of graph do you think the psychologists should draw? Use technology or draw one by hand.

b)   Describe the relationship shown.

	A	B	C	D	E	F	G	H	I	J	K	L	M	N	O	P	Q	R
1	Age	16	20	25	12	18	40	14	44	21	29	45	55	52	61	43	57	16
2	Items	12	15	14	11	16	8	13	7	15	17	9	8	7	18	6	5	18

5   Katie makes a summary note to show what each type of graph or chart displays. Complete the last column of her note, choosing from: scatter graph, compound bar graph, pie chart and compound line graph.

The graph displays...	Type of graph or chart
a total amount shared into different categories.	
the frequency of different categories with clusters or stacks of bars showing groups in each category.	
trends in data over a continuous numerical scale with a key to identify groups.	
possible links between two numerical data sets.	

6   In Exercise 12B, 12C and 12E questions 1 and 2 asked you to draw a chart by hand. If you have access to technology go back to these questions, type in the data sets and use the computer to produce the answers. Compare these to the ones in your jotter. You might be surprised by the ones you like the best. You can also try Exercise 12F questions 2, 3 and 4.

Check-up 👍

1 Copy the table. List each data set under the correct heading.

eye colour, price, hobby, flight time, exam subject, exam length

Categorical	Numerical

2 David's phone collects and displays data on his screen time. He is trying to cut down on his phone use and buys a book of maths puzzles to give his brain a workout in week 2.

a) How many hours did David spend using apps in week 1?

b) How many hours did he spend on his phone altogether in week 1?

c) Find the percentage drop in his gaming from week 1 to 2.

d) Do you think David's plan to cut down his screen time worked? Give evidence for your answer.

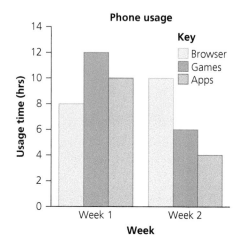

3 Prithika has an app on her phone which tracks how many steps she walks every day.

The app displays the graph shown.

a) How many steps did Prithika do on Thursday morning?

b) How many steps did she do on Thursday afternoon?

c) What fraction of her Thursday steps were taken in the morning? Give your answer in its simplest form.

d) Describe 2 ways the data for Saturday is different to Prithika's weekday pattern.

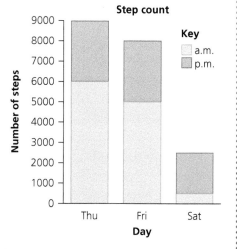

4 The manager of a party venue records data from two parties. She counted the number of people on the dancefloor every half hour each night. The results are shown.

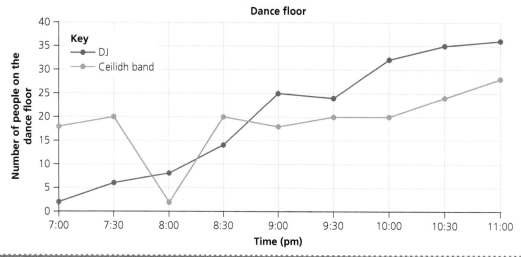

a) How many people were dancing to the ceilidh band at 7·30 pm?

b) Describe the trend for the ceilidh band. What happened at 8 pm?

c) How many people were dancing to the DJ at 8:30 pm?

d) Describe the trend for the DJ.

e) How many more people were dancing to the ceilidh band than the DJ at 8:30 pm?

f) Which entertainment would you say gets more people dancing? Give evidence from the graph for your answer.

5 Ms Harper is a driving instructor. She has a total of 60 clients and creates the pie chart shown to display their ages.

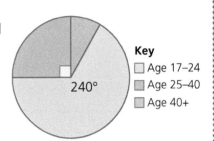

Key
☐ Age 17–24
▨ Age 25–40
▨ Age 40+

240°

a) What fraction of her clients are aged 25–40?

b) How many clients are aged 25–40?

c) What fraction of her clients are aged 17–24?

d) How many clients are aged 17–24?

e) How many of the people learning to drive are over 40?

6 The spreadsheet shows typical monthly temperatures in Crieff Scotland and Sydney Australia.

	A	B	C	D	E	F	G	H	I	J	K	L	M
		Jan	Feb	March	April	May	June	July	Aug	Sept	Oct	Nov	Dec
1													
2	Crieff °C	7	7	8	10	14	16	18	17	15	12	9	7
3	Sydney °C	27	27	26	23	20	18	17	19	21	23	24	26

a) Create a compound line graph to display this data.

Show the month of the year on the x-axis and the temperature on the y-axis. Include colour and a key.

If you have access to technology your teacher may ask you to use this. Use the options to add a title and axes labels.

b) What is the range (highest – lowest) of temperatures in Crieff over the year?

c) What is the range of temperatures in Sydney?

d) How different is the temperature in Sydney compared to Crieff? Give evidence from the graph.

e) In which month is the temperature in Crieff and Sydney most similar?

7 A newspaper headline says 'Eating Bananas leads to Exam Success'.

'We interviewed 10 pupils who were happy with their exam results, 8 of them said they liked bananas.'

What do you think of the headline? What advice would you give the newspaper editor?

13 Angles

▶ Naming angles

An **angle** is formed when two straight lines meet at a single point (a **vertex**). To name a given angle:

- Use three letters.
- The middle letter must be the vertex where the angle is formed.
- Use the angle symbol ($\angle$). Note: throughout this chapter angles are measured in degrees. Make sure to include units in your final answers!

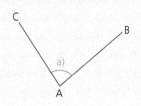
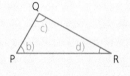
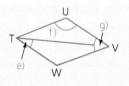

a) $\angle$CAB or $\angle$BAC

b) $\angle$QPR or $\angle$RPQ
c) $\angle$PQR or $\angle$RQP
d) $\angle$PRQ or $\angle$QRP

e) $\angle$WTV or $\angle$VTW
f) $\angle$TUV or $\angle$VUT
g) $\angle$WVU or $\angle$UVW

Exercise 13A

1   Write down two possible names for each shaded angle.

a)   **b)**   **c)**   **d)**

In practice, we only give one of the possible names for a given angle. For the remainder of this chapter, when asked to name an angle you need only do so in one way.

2   In each diagram, name each shaded angle and state whether it is acute, right, obtuse or reflex.

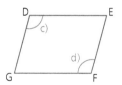

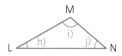
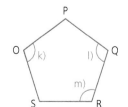
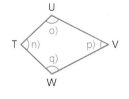

▶ Vertically opposite angles

When two straight lines cross over, they create four angles.

Angles opposite each other in this way are equal.

We call these vertically opposite angles. Remember:

- Angles that form a straight line add up to 180°.
- A pair of angles that add up to 180° are supplementary angles.
- Angles that form a right angle add up to 90°.
- A pair of angles that add up to 90° are complementary angles.

Worked examples

1 Name the angle that is vertically opposite to ∠ABE.

2 Calculate the size of each unknown angle.

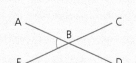

∠CBD or ∠DBC.

∠FJG is vertically opposite ∠IJH, so ∠FJG = 126°.

∠FJI and ∠IJH form a straight angle.

$$\begin{array}{r} 71 \\ \cancel{180} \\ -126 \\ \hline 54 \end{array}$$

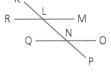

∠FJI = 54°

∠GJH is vertically opposite ∠FJI, so ∠GJH = 54°.

Exercise 13B

1 Copy and complete these statements equating pairs of vertically opposite angles.

a)

∠ABC = ∠EBD
∠ABE = . . .

b)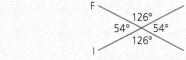

∠GJH = . . .
∠GJF = . . .

c)

∠RLN = . . .
∠ONP = . . .

2 Copy the diagrams, filling in the size of each unknown angle.

a)

b)

c)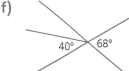

d)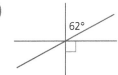

e)

f)

▶ Corresponding angles

When a pair of **parallel lines** is crossed by a third line, **corresponding angles** are formed. To identify corresponding angles:

- Look for a pair of parallel lines connected by a third line.
- Identify an 'F' shape: the angles under the 'arms' of the F are corresponding and therefore equal.

Transverse line

Parallel lines

Examples of corresponding angles are marked on the diagrams.

Worked examples

1 In the diagram, which angle is corresponding to

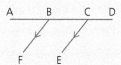

   a) ∠ABF   ∠BCE is corresponding to ∠ABF
   b) ∠ECD   ∠FBC is corresponding to ∠ECD.

2 In the diagram, find the size of

   a) ∠GBC   ∠GBC is corresponding to ∠EGD so ∠GBC = 107°
   b) ∠FGB   ∠FGB is vertically opposite ∠EGD so ∠FGB = 107°
   c) ∠BGD   ∠BGD and ∠EGD form a straight angle

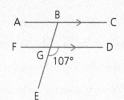

$$\begin{array}{r} \overset{7}{1}\\ 180\\ -107\\ \hline 73 \end{array}$$

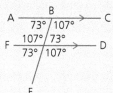

∠BGD = 73°

Exercise 13C

1 For each diagram, name the pair of corresponding angles.

   a) 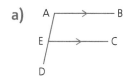   b) 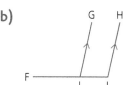   c)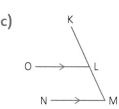

2 Copy each diagram, marking corresponding angles with the symbols shown.

   a)   b) 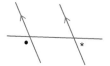   c)

3 Copy each diagram, filling in all unknown angles.

a)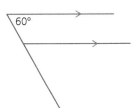
60°

b)
130°

c)
23°

d)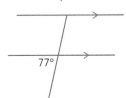
77°

e)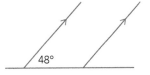
48°

f)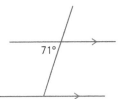
71°

g)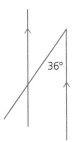
36°

h)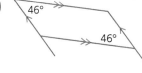
46°
46°

i)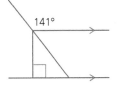
141°

4 Look at the diagram. Explain how you know lines AB and CD are **not** parallel.

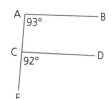
A 93° B
C 92° D
E

5 Copy the triangle diagram.

a) Calculate the size of each unknown angle.

b) Look at the angles in the large and small triangles. Comment on anything you notice.

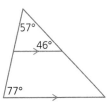
57°
46°
77°

6 Form an equation and solve it to find the value of *a*.

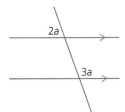
2a
3a

7 A manufacturer designs a wall bracket to hold shelves. In order to pass quality control, the two arms of the bracket must be parallel. Will this one pass quality control? Give a reason for your answer.

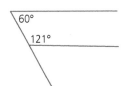
60°
121°

▶ Alternate angles

Alternate angles may be formed when parallel lines are connected. To identify alternate angles:

- Look for a pair of parallel lines connected by a third line.
- Identify a 'Z' shape: the angles in the corners of the Z are alternate and therefore equal.

Two pairs of alternate angles are marked on the diagram.

Worked examples

1 In the diagram, which angle is alternate to:

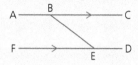

a) ∠BEF

∠EBC is alternate to ∠BEF

b) ∠DEB?

∠ABE is alternate to ∠DEB

2 In the diagram, find the size of:

a) ∠KLI

b) ∠GLI

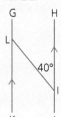

∠KLI is alternate to ∠LIH so ∠KLI = 40°

∠GLI and ∠KLI form a straight angle

$$\begin{array}{r} 180 \\ -40 \\ \hline 140 \end{array}$$

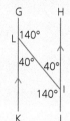

∠GLI = 140°

Exercise 12D

1 For each diagram, name the pair of alternate angles.

a)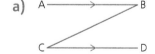

b)

c)

2 Copy each diagram, marking alternate angles with the symbols given.

a)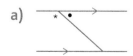

b)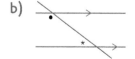

c)

3 Copy the diagrams, filling in all unknown angles.

a)
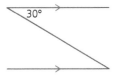

b)
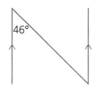

c)
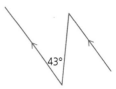

d)
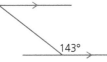

e)
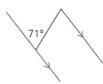

f)
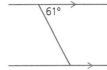

g)

h)

i)
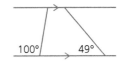

Worked example For the diagram shown:

a) Form an equation
b) Solve it to find the value of x
c) Calculate the size of each angle.

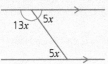

a) Spotting alternate angles, we can see that $5x + 13x = 180$

b) $5x + 13x = 180$
$18x = 180$
$\div 18 \quad \div 18$
$x = 10°$

c) Substituting gives $5x = 50°$ and $13x = 130°$

---

4 For each of the diagrams:

i) Form an equation   ii) Solve it to find the value of x iii) Calculate the size of all the angles.

a)

b)
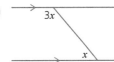

c)
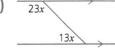

5 Say which of these lines are parallel and explain how you know.

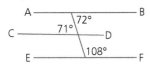

▶ Angles in a triangle

The angles inside any triangle add up to 180°. We can prove this using what we already know about alternate angles.

First, we draw a general triangle and label the angles *a*, *b* and *c*.

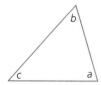

Next, we add a straight line parallel to one side of the triangle.

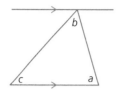

Finally, we identify two pairs of alternate angles. Since *a*, *b* and *c* form a straight angle they clearly add up to 180°, and since ours was a general triangle this will be true of **any** triangle.

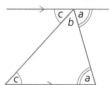

To find a missing angle in a triangle:

● Add up the known angles.

● Subtract the answer from 180°.

Worked example   In each triangle, calculate *x*.

a)

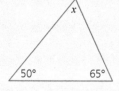

$$\begin{array}{r}71\\50\\+_165\\\hline115\end{array}\quad\begin{array}{r}1\,80\\-115\\\hline65\end{array}$$

(Note: remember to include degrees in your answer)

x = 65°

b)

(Note that there is a right angle)

$$\begin{array}{r}71\\90\\+_131\\\hline121\end{array}\quad\begin{array}{r}1\,80\\-121\\\hline59\end{array}$$

x = 59°

c)

(Note that this triangle is isosceles, so two angles are equal)

$$\begin{array}{r}71\\54\\+\;54\\\hline108\end{array}\quad\begin{array}{r}1\,80\\-108\\\hline72\end{array}$$

x = 72°

Exercise 13E

1 Calculate *x* in each triangle.

a)

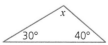

b)

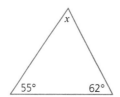

c)

d) 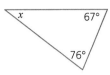 67° 76° x

e) 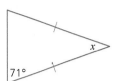 x 43°

f) 71° x

g)  122° x

h) 41° 39° x

i)  x

2 Calculate the size of each unknown angle in these diagrams.

a) 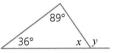 89° 36° x y

b) 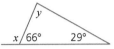 y x 66° 29°

c) y x z 51°

d) 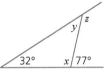 y z 32° x 77°

e) 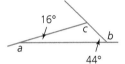 16° c b a 44°

f) 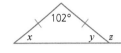 102° x y z

g) 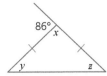 86° x y z

h) 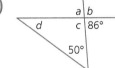 a b d c 86° 50°

i) 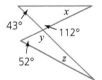 43° x y 112° 52° z

3 For each triangle:
  i) form an equation
  ii) solve it to find the value of *x*
  iii) calculate the size of all the angles.

a) 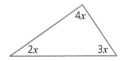 4x 2x 3x

b)  5x 2x 5x

c) 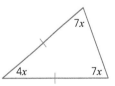 7x 4x 7x

d) x x x

e)  10x 4x 16x

f) 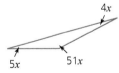 4x 5x 51x

4 Two of the angles in a triangle are 46° and 58°. Is the triangle isosceles? Explain your answer.

5 An isosceles triangle ABC is right-angled at ∠BCA. Write down the sizes of angles ∠CBA and ∠CAB.

6 Explain why a triangle cannot have more than one obtuse angle.

7 Copy the diagram, filling in the size of each angle.

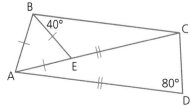

▶ A mixture!

When calculating angles within diagrams, we must look for all the patterns we have already considered, i.e.:

- angles that form a straight line sum to 180°
- angles that form a right angle sum to 90°
- angles that form a full turn sum to 360°
- vertically opposite angles are equal.

- corresponding angles are equal
- alternate angles are equal
- angles in a triangle sum to 180°

Worked example

Calculate the size of each unknown angle in these diagrams.

a)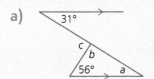

$a = 31°$ (alternate angles)

a, b and 56° are angles in a triangle

$$\begin{array}{r} 56 \\ + 31 \\ \hline 87 \end{array} \qquad \begin{array}{r} {}^{1}7^{1} \\ \cancel{1}80 \\ - 87 \\ \hline 93 \end{array}$$

$b = 93°$

b and c are supplementary angles

$$\begin{array}{r} {}^{1}7^{1} \\ \cancel{1}80 \\ - 93 \\ \hline 87 \end{array}$$

$c = 87°$

b)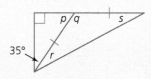

$k = 81°$ (vertically opposite angles)

m and k are supplementary angles

$$\begin{array}{r} {}^{1}7^{1} \\ \cancel{1}80 \\ - 81 \\ \hline 99 \end{array}$$

$m = 99°$

$l = 99°$ (vertically opposite angles)

n and m are corresponding angles

$n = 99°$

n and o form a full turn

$$\begin{array}{r} {}^{2}{}^{1}5^{1} \\ \cancel{3}\cancel{6}0 \\ - 99 \\ \hline 261 \end{array}$$

$o = 261°$

c)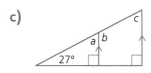

$$\begin{array}{r} 90 \\ + {}_{1}35 \\ \hline 125 \end{array} \qquad \begin{array}{r} {}^{7}1 \\ \cancel{1}80 \\ - 125 \\ \hline 55 \end{array}$$

$p = 55°$

p and q are supplementary angles

$$\begin{array}{r} {}^{7}1 \\ \cancel{1}80 \\ - 55 \\ \hline 125 \end{array}$$

$q = 125°$

Note that the triangle on the right is isosceles.

$$\begin{array}{r} {}^{7}1 \\ \cancel{1}80 \\ - 125 \\ \hline 55 \end{array} \qquad \begin{array}{r} 2\,7 \cdot 5 \\ 2\,\overline{)5\,{}^{1}5 \cdot {}^{1}0} \end{array}$$

$r = 27 \cdot 5°$

$s = 27 \cdot 5°$

Exercise 13F

1 Calculate the size of each marked angle in these diagrams.

a)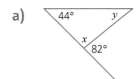

b)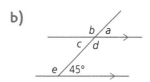

c)

d)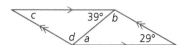

e)

f)

g)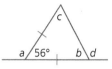

h)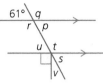

i)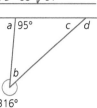

j)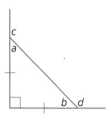

k)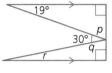

l)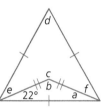

m)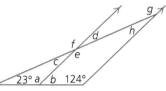

n)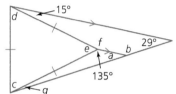

o)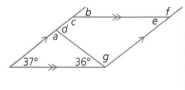

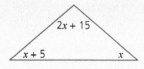

Worked example   Form an equation and solve it to find the value of x.

Angles in a triangle sum to 180° so

$2x + 15 + x + 5 + x = 180$

Collecting like terms gives

$4x + 20 = 180$

We now solve as normal

$4x + 20 = 180$
$\quad - 20 \quad - 20$
$\quad\quad 4x = 160$
$\quad\quad \div 4 \quad \div 4$
$\quad\quad\quad x = 40$

So our solution is $x = 40°$

---

2   For each diagram, form an equation and solve it to find the value of x.

a)

b)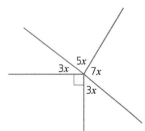

c)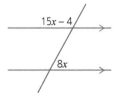

d)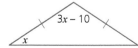

e)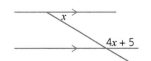

f)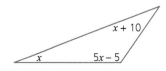

▶ Angles in polygons

Every quadrilateral can be split into two triangles.

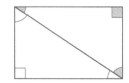 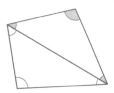

Since the angles in a triangle sum to 180°, this means the angles in a quadrilateral sum to 360°.

● Angles in a quadrilateral sum to 360°.

● To find a missing angle, add known angles and subtract from 360°.

Worked example Calculate the size of the unknown angles in these quadrilaterals.

a)
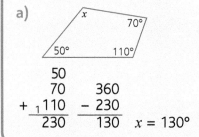

$$
\begin{array}{ll}
50 & \\
70 & 360 \\
+_1 110 & -230 \\
\hline
230 & 130 \quad x = 130°
\end{array}
$$

b)
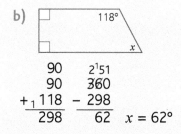

$$
\begin{array}{ll}
90 & 2^1 51 \\
90 & 3\cancel{6}0 \\
+_1 118 & -298 \\
\hline
298 & 62 \quad x = 62°
\end{array}
$$

c)
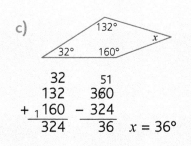

$$
\begin{array}{ll}
32 & 51 \\
132 & 3\cancel{6}0 \\
+_1 160 & -324 \\
\hline
324 & 36 \quad x = 36°
\end{array}
$$

Exercise 13G

1 Calculate the size of x in each quadrilateral.

a)
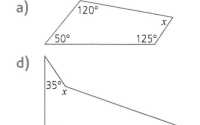

b)
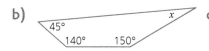

c)
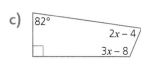

d)

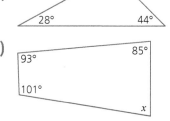

e)

f)

2 For each of the quadrilaterals, form an equation and solve it to find the value of x.

a)
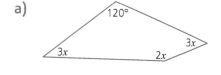

b)
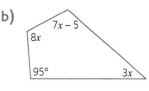

c)

A regular shape is one in which all the interior angles are equal and all edges are the same length. We can calculate the interior angles of a regular shape by the following method.

● Draw the shape and split into isosceles triangles.

● Calculate the size of the central angles (360 ÷ the number of sides).

● Calculate the size of the remaining angles in the triangles.

● Double this to find the size of an interior angle.

Worked example

Calculate the sizes of the interior and exterior angles of a regular pentagon.

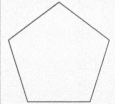 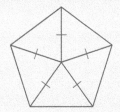 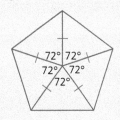 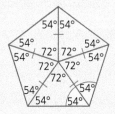

The central angles will be given by **360 ÷ 5 = 72°**.

As the triangle is isosceles, the remaining angles are given by $(180 - 72) \div 2$.

An interior angle will then be given by 54×2.

The interior angle is 108°.

$$\begin{array}{r} 71 \\ 180 \\ -\ 72 \\ \hline 108 \end{array}$$

$$5\overline{)3^36^10}\quad 72$$

$$2\overline{)1^108}\quad 54$$

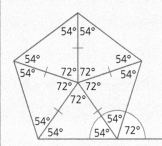

To calculate the size of an exterior angle, we simply subtract the interior angle from 180°.

In this case an exterior angle is given by $180 - 108$.

$$\begin{array}{r} 71 \\ 180 \\ -\ 108 \\ \hline 72 \end{array}$$

The exterior angle is 72°.

Exercise 13H

1   Here is a regular hexagon.
    a) Calculate the size of each central angle x.
    b) Hence, calculate the size of the remaining angles, y, in each isosceles triangle.
    c) Write down the size of the interior angles of the regular hexagon.
    d) Write down the size of the exterior angles of the regular hexagon.

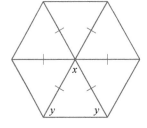

2   Follow the steps from the previous question to calculate the size of the interior and exterior angles of these regular shapes.

a) Octagon (8 sides)   b) Decagon (10 sides)   c) Icosagon (20 sides)   d) Dodecagon (12 sides)

e) Nonagon (9 sides)   f) Octadecagon (18 sides)   g) Hectogon (100 sides)   h) Heptagon (7 sides), giving your answer to 2 decimal places

▶ Constructions

With a ruler, a protractor and compasses we can create accurate drawings of polygons.

Worked example Make an accurate drawing of the triangle sketched here.

- Use your ruler to draw a side of length 5 cm horizontally.
- Place the centre of the protractor at one end of the line, measure the angle of 40° and make a faint mark.
- Place your ruler so that the 'zero' is at the end of the line and the ruler passes through the mark.
- Keeping the ruler in place, draw a line of the length 4 cm.
- Use your ruler to join the endpoints of the two lines, forming the final side.

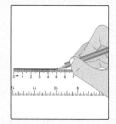   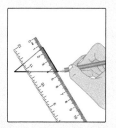 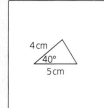

Exercise 13I

1 Make an accurate drawing of the triangle sketched here.

2 Make an accurate drawing of the triangle sketched here.

3 Draw a triangle with sides 5·5 cm and 2·8 cm and an included angle of 76°.

4 Draw a triangle with sides 3 cm, 3·5 cm and an included angle of 123°.

5 Draw a triangle with sides 4·1 cm, 3·7 cm and an included angle of 95°.

Worked example

Make an accurate drawing of the triangle sketched here.

- Use your ruler to draw a side of length 6 cm horizontally.
- Place the centre of the protractor at one end of the line, measure an angle of 35° and make a mark. Then, using your ruler, draw a faint line from the end of the given line through this mark.

- Place the centre of the protractor at the opposite end of the line, measure an angle of 60° and make a mark. Use your ruler to draw a faint line from the end of the first line, through this mark, ensuring that it crosses the previous line.

- From the point at which the two faint lines intersect, use your ruler to draw in the two sides. Rub out any faint lines that remain.

 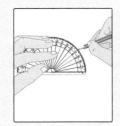  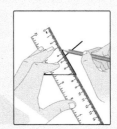 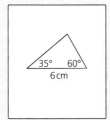

Exercise 13J

1 Make an accurate drawing of the triangle sketched here.

2 Make an accurate drawing of the triangle sketched here.

3 Draw a triangle with a side of length 4 cm and two angles of 50°.

4 Draw a triangle with a side of 5 cm and angles 105° and 30°.

5 Draw a triangle with a side of 6·4 cm and angles 140° and 28°.

Worked example

Make an accurate drawing of the triangle sketched here.

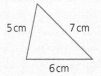

- Draw one side using a ruler.

- Set your compasses to the required length for the second side, place the point at one end of your line and draw a faint arc.

- Set your compasses to the required length for the final side, place the point at the other end of the line and draw a second faint arc.

- Using your ruler, join the point at which the two arcs intersect to each end of the line.

 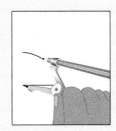 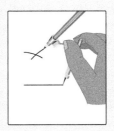 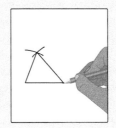 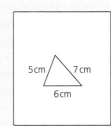

Exercise 13K

1 Make an accurate drawing of the triangle sketched here.

2 Make an accurate drawing of the triangle sketched here.

3 Draw a triangle with sides of length 5·2 cm, 4·7 cm and 3·5 cm.

4 Draw a triangle with sides of length 7 cm, 6·2 cm and 4·7 cm.

5 Draw a triangle with sides of length 8·3 cm, 4·8 cm and 6 cm.

6 Explain why you cannot construct a triangle with sides of length 6 cm, 2 cm and 3 cm.

Worked example

Construct a regular pentagon with sides of length 4 cm.

- Calculate the size of the interior angles using the method previously shown.
- Draw a side of length 4 cm horizontally.
- Place the centre of the protractor at one end of the line, measure an angle of 108° and make a mark.
- Align the ruler from the end of the line through the mark and draw the second side of length 4 cm.
- Start at the end of this new line and repeat this process until the polygon is complete.

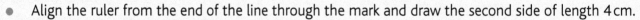

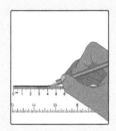

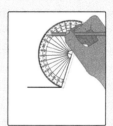
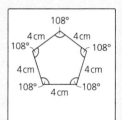

Exercise 13L

1 Construct an equilateral triangle with sides of length 4·5 cm.

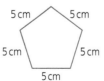

2 Construct a regular pentagon with sides of length 5 cm.

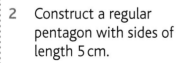

3 Construct a regular hexagon with sides of length 3 cm.

4 Construct a regular heptagon (7-sided shape) with sides of length 3·5 cm.

5 Construct a regular octagon with sides of length 3 cm.

▶ Bearings

Bearings describe the position of objects relative to each other and are crucial for navigation. When measuring and writing bearings, there are three key points to remember.

- Start measuring from north.
- Measure in a clockwise direction.
- Always use three figures to write down bearings.

Worked example For each diagram, measure the bearing of B from A.

a)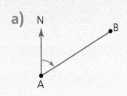

We must place our protractor so that the centre is over point A and the zero line points north.

Reading clockwise we have an angle of 55° but remember we use three figures for bearings.

The bearing of B from A is 055°.

b)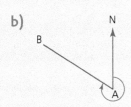

As this bearing is greater than 180°, we must place our protractor facing to the left of north.

Reading clockwise from zero (South), our protractor reads 110°, so the bearing of B from A is

```
   180
 + 110
 ─────
   290°
```

Exercise 13M

1. Draw an eight-point compass and write down the bearing of each of the main compass points.

2. Write down the three-figure bearing of:

   a) east   b) west   c) south   d) north-east

   e) south-west   f) north   g) south-east   h) north-west

3. For each of these diagrams, measure and write down the bearing of B from A.

   a)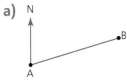

   b)

   c)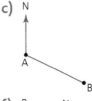

   d)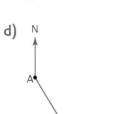

   e)

   f)

➡

Worked example If the bearing of B from A is 030°, what is the bearing of A from B?

We can see this clearly by drawing a diagram.
By extending the line from A to B and adding north lines, we have a pair of corresponding angles.

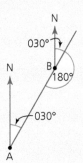

$$\begin{array}{r} 30 \\ +_1180 \\ \hline 210° \end{array}$$

The bearing of A from B is therefore 210°.

---

4   For each of these diagrams, calculate the bearing of A from B.

a)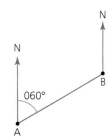

b)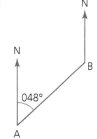

c)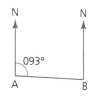

d)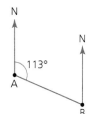

e)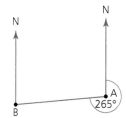

f)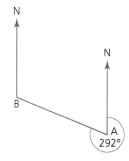

5   Calculate the bearing of A from B if the bearing of B from A is

   a) 040°   b) 035°   c) 104°   d) 256°

   e) 300°   f) 017°   g) 234°   h) 108°

6   Using a ruler and protractor, draw diagrams like the ones in the question 4 such that the bearing of B from A is

   a) 050°   b) 070°   c) 120°   d) 175°

   e) 296°   f) 193°   g) 351°   h) 007°

7   For each of your diagrams for question 6, calculate the bearing of A from B and then use your diagram to verify.

Worked example

Edinburgh is approximately 42 miles from Glasgow on a bearing of 082°. Make a scale drawing to show the relative positions of Edinburgh and Glasgow using a scale of 1 cm : 10 miles.

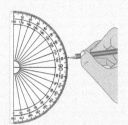

Draw a north line and measure the bearing of 082°

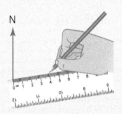

Use the scale to calculate and draw a line of the required length.
42 ÷ 10 = 4·2

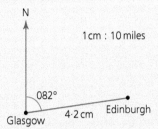

Label diagram and show the scale.

8   Campbeltown is approximately 48 miles from Kilmarnock on a bearing of 276°. Make a scale drawing to show the relative positions of Kilmarnock and Campbeltown using a scale of 1 cm : 5 miles.

9   It is approximately 970 km from John O'Groats to Land's End on a bearing of 190°. Make a scale drawing using a scale of 1 cm : 100 km that shows the relative positions of John O'Groats and Land's End.

10   A ship sets off from port and sails due East to point A 50 km away. It then changes course and sails for 80 km to point B on a bearing of 040°. The ship then sails from point B directly back to port.

    a)   Use a suitable scale and make a scale drawing showing the ship's journey.

    b)   Use your scale drawing to estimate the distance from point B to port.

    c)   What is the bearing of port from point B?

11   A plane takes off from an airport and flies 800 km on a bearing of 130°. It lands for refuelling before taking off once more and flying 450 m on a bearing of 240°. Use a scale drawing to determine how far the plane is from the airport where it first took off.

12   Barry has to set out markers for an orienteering course. He sets off from camp and walks 2 km on a bearing of 170° then places marker A. From this point he walks 4 km due East and places marker B. Barry then walks in a straight line back to camp.

    a)   Construct a scale drawing with an appropriate scale to show Barry's journey.

    b)   Use your scale drawing to estimate the distance from marker B back to camp.

    c)   Use your scale drawing to measure the bearing of marker B from camp.

13   Daanyaal, Maya and Tom all set off from the same point running in different directions. Daanyaal runs 20 m South, Maya runs 30 m on a bearing of 120° and Tom runs 20 m on a bearing of 240°.

    a)   Construct a scale drawing with an appropriate scale to show the relative positions of Daanyaal, Maya and Tom.

      Use your scale drawing to estimate

    b)   the distance between Daanyaal and Maya

    c)   the distance between Maya and Tom

    d)   the distance between Tom and Daanyaal.

Check-up 👍

1   For each diagram, write two possible names for each shaded angle.

a)

b)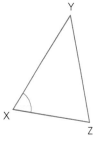

c)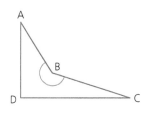

d)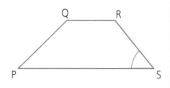

2   Copy and complete these sentences.

a)   Angles in a triangle sum to _____.

b)   Angles around a point sum to _____.

c)   Angles in a quadrilateral sum to _____.

d)   Angles that form a 'Z' shape are called _____ angles.

e)   Angles that form an 'F' shape are called _____ angles.

3   For each shape, write down the size of the marked angles.

a)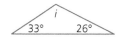

b)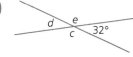

c)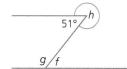

d)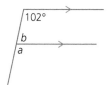

e)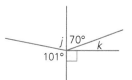

f)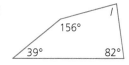

4   For each diagram, form an equation, solve it to find x and calculate the size of each angle.

a)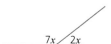

b)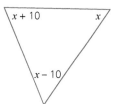

c)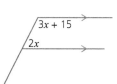

d)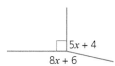

e)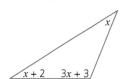

f)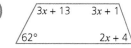

5   Write down the size of the marked angles in each of the diagrams.

a)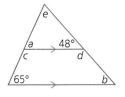

b)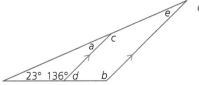

c)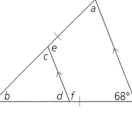

d)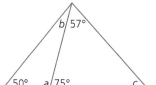

e)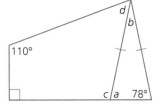

f)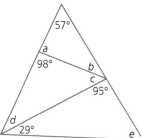

6   Draw a triangle with sides of length 7 cm and 5·5 cm and an included angle of 65°.

7   Draw a triangle with a side of length 5 cm and angles of 35° and 100°.

8   Draw a triangle with sides of length 12 cm, 8·5 cm and 6 cm.

9   Draw a regular pentagon with sides of length 5 cm.

10   Draw an equilateral triangle with sides of length 6·5 cm.

11   Calculate the size of the interior and exterior angles of the following regular shapes.

   a) Hexagon   b) Octagon   c) Pentadecagon (15 sides)

12   Which compass point are you facing if you look north and turn to a bearing of:

   a) 090°   b) 135°   c) 315°   d) 045°?

13   In each of the diagrams, measure the bearing of B from A.

a)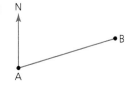

b)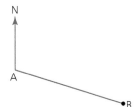

c)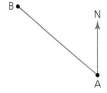

14   Berlin is approximately 880 km from Paris on a bearing of 059°. Choose a suitable scale and make a scale drawing showing the relative positions of Paris and Berlin.

15   Paisley is approximately 87 km from Inverness on a bearing of 178°. Use a scale of 1 cm : 10 km to make a scale drawing showing the relative positions of Inverness and Paisley.

16   A cruise ship sets off from port and sails 50 km south-east where it docks at an island to let off passengers. From there it sets off on a bearing of 025° and sails for 70 m to a marina. After a short stay the ship heads directly back to port.

   a)   Make a scale drawing to show the journey of the cruise ship.

   b)   Use your drawing to determine the length from the marina to port.

   c)   If the ship sails at an average speed of 35 km/h, how long did the entire journey take (not including stops)?

14 Probability

► Probability scale and likelihood

We say that an **event** is a particular thing that may or may not occur, and that the **probability** of a given event is the likelihood with which it will occur. Probability ranges from **impossible** (something that cannot possibly occur) to **certain** (something that must occur). We can show the **relative probabilities** of a set of events by placing them on a probability scale.

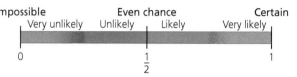

If an event is impossible, it has zero chance of happening (its probability is 0). If, on the other hand, an event will certainly happen, we say its probability is 1. If there is exactly the same chance of an event occurring or not occurring, then its probability is one half (**even chance** or **fifty-fifty**). Note that this is much more specific than simply saying that it might or might not occur.

Throughout this chapter, we will refer to a deck of cards. You should be familiar with these facts.

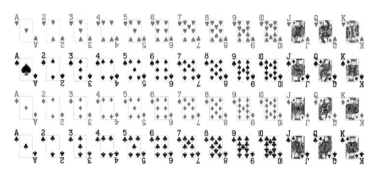

- A deck has 52 cards in total.
- There are four suits: hearts, diamonds, clubs and spades.
- Hearts and diamonds are red. Clubs and spades are black.
- Each suit contains 13 cards: 10 numbered cards (ace to 10) and three face cards: jack, queen and king.
- Throughout this chapter, we regard aces as the lowest value and kings as the highest value.

Exercise 14A

1   For each event, write down the phrase from the probability scale that best describes its likelihood.

   a) Turning over a red card from a shuffled deck
   b) Tossing a coin and getting 'heads' ten times in a row
   c) Getting homework today
   d) The day after Saturday being Sunday
   e) Turning over a club from a shuffled deck of cards
   f) Getting a draw in a single game of 'rock, paper, scissors'

2   A sock drawer contains 4 grey, 10 white and 18 black pairs of socks.

   a)   How many individual socks must someone remove to be certain of having a matching pair?

   b)   Describe the likelihood that a sock removed at random is grey.

   c)   How many new pairs of coloured socks must be added to the drawer to give an even chance that a randomly selected sock is black?

3   Look at the spinner and write down words from your probability line that describe the likelihood of the spinner landing on a

   a) 2   b) 1   c) 3   d) 4   e) prime number   f) square number

▶ Calculating probability

In order to calculate the probability of a given event occurring:

- Count how many ways we can get the desired result (favourable outcomes).
- Count how many things could happen (possible outcomes).
- Use the formula: probability $= \dfrac{\text{number of favourable outcomes}}{\text{number of possible outcomes}}$.

Remember: always simplify fractions when possible.

Worked examples

1 Calculate the probability that a card picked at random from a shuffled deck is a queen.

Remember that there are 4 queens in 52 cards, so

$$P(\text{queen}) = \frac{4}{52}$$
$$= \frac{1}{13}$$

3 The probability of winning a prize in a competition is $\frac{2}{15}$. What is the probability of **not** winning a prize?

The probability that an event does not happen is one minus the probability that it does happen, so

$$P(\text{not winning}) = 1 - \frac{2}{15}$$
$$= \frac{15}{15} - \frac{2}{15}$$
$$= \frac{13}{15}$$

2 A bag contains 6 green marbles, 3 red marbles and 5 blue marbles.

a) Calculate the probability that a marble drawn at random is green.

There are 6 green marbles out of a total of 14 marbles, so

$$P(\text{green}) = \frac{6}{14}$$
$$= \frac{3}{7}$$

b) A red marble is drawn from the bag and not replaced. Calculate the probability that the next marble drawn from the bag is also red.

Once a red marble has been drawn and not replaced there are now 2 red marbles left out of a total of 13 marbles, so

$$P(\text{red}) = \frac{2}{13}$$

Exercise 14B

1 A bag contains 4 red balls and 8 yellow balls. If a ball is removed at random what is the probability that it is: **a)** red **b)** yellow **c)** green?

2 A box contains 15 red pens and 10 blue pens. A pen is removed at random.

a) What is the probability that it is red? b) What is the probability that it is blue?

c) A pen is removed at random and **not replaced**. If the pen was blue, what is the probability that the next pen removed is also blue?

→

3   Tiles with the letters of the word 'MATHEMATICS' are placed in a bag. If a tile is drawn at random, what is the probability that it is:

     a)   an A   **b)**   a consonant   **c)**   a vowel?

4   A deck of cards is shuffled and the top card is revealed. What is the probability that the top card is:

     a)   a club   **b)**   an ace   **c)**   a face card   **d)**   a red number card

     e)   a number card   **f)**   a black 8   **g)**   a prime number   **h)**   a square number?
        greater than 5

5   On the spinner shown, what is the probability of getting:

     a)   an 'A'   **b)**   a vowel   **c)**   a consonant

     d)   a letter that can be 'coloured in'   **e)**   a letter in the second half of the alphabet?

6   What is the probability of a given event **not** happening if the probability that it happens is:

     a) $\dfrac{1}{2}$ **b)** $\dfrac{2}{5}$ **c)** $\dfrac{4}{13}$ **d)**   0·7   **e)**   0·15   **f)**   1?

7   A box contains 6 red, 3 green, 4 blue, 5 orange and 6 yellow lollipops. What is the probability that a lollipop taken at random is:

     a)   green   **b)**   yellow   **c)**   blue or orange   **d)**   not red

     e)   red or yellow   **f)**   not green   **g)**   red, green or orange   **h)**   not blue or green?

8   Nicky's MP3 player has 200 rock songs, 40 country songs, 350 pop songs and 110 heavy metal songs. She puts the player on shuffle. What is the probability that the first song is:

     a)   a country song   **b)**   not a rock song   **c)**   a pop or heavy metal song?

     Once a song is played, it is not repeated. The first ten songs played at random are all pop songs.

     What is the probability that the next song is:

     d)   also a pop song   **e)**   not a pop song   **f)**   a rock song?

9   This is a set of shape cards for a maths game.

     What is the probability that a card selected at random has on it:

     a)   a quadrilateral

     b)   a shape with a curved edge

     c)   a polygon with an angle sum greater than 180°

     d)   a shape with no lines of symmetry?

10   Kirsten and Billy invent a game to play in maths. They have cards numbered 1 to 25 and take turns to turn two cards over. If the two cards share a particular property, the player gets to keep those two. Otherwise, both cards are discarded. The first card Kirsten turns over is the 4. What is the probability that her next card is also

     a)   a multiple of 4   **b)**   a square number   **c)**   an even number?

     Kirsten's second card was the 5. Billy now takes his turn and turns over the 11. What is the probability that Billy's second card is:

     d)   a prime number   **e)**   an odd number   **f)**   a multiple of 5?

11 Brian plays a game involving rolling a 12-sided die. What is the probability that he rolls:

    a) a prime number   **b)** a factor of 12   **c)** a triangular number?

12 Katy takes a survey of her classmates' shoe sizes and produces this chart.

What is the probability that a child chosen from the class at random has:

    a) size 6 feet

    b) size 8 feet

    c) size 2 feet

    d) size 4 or 5 feet

    e) feet smaller than size 5

    f) feet larger than size 4?

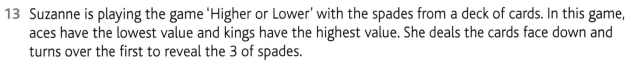

Shoe sizes in S1 Class

13 Suzanne is playing the game 'Higher or Lower' with the spades from a deck of cards. In this game, aces have the lowest value and kings have the highest value. She deals the cards face down and turns over the first to reveal the 3 of spades.

    a) What is the probability that the next card is higher?

    b) Suzanne guesses 'higher' and turns over the next card to reveal the 9 of spades. Is the next card more likely to be higher or lower?

14 The colours of cars in a multi-storey car park are shown in this chart.

What is the probability that the next car to leave the car park is

    a) black   **b)** yellow

    c) red or green   **d)** white or silver

    e) not blue?

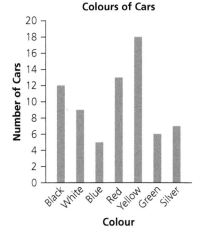

Colours of Cars

15 Copy and complete this table showing the possible totals when rolling two dice, then use it to answer the questions. The first row has been started for you.

    a) What total are you most likely to achieve when rolling two dice?

    b) How many ways are there to roll a 2-digit total?

    c) What is the probability of rolling a double?

    d) What is the probability of rolling an odd total?

    e) What is the probability of rolling a total less than 5?

		Die 1					
		1	2	3	4	5	6
	1	2	3	4	5	6	
	2						
	3						
Die 2	4						
	5						
	6						

▶ Using probability

As well as calculating the probability of an event happening or not happening, we may wish to calculate how many times we would expect that event to occur over a number of trials. This is the theoretical probability, and it is important to realise that we may observe different results in practice. To work this out:

- Calculate the probability of the event occurring.
- Multiply the probability by the number of trials.

We may also wish to compare two probabilities to identify which is more likely. To do so:

- Write both probabilities as percentages, decimals or fractions with a common denominator, using a calculator if necessary.
- The one with the larger probability is the one that is more likely to occur.

Worked examples

1   How many times would you expect to roll a one on a regular die if you tried 90 times?

First, calculate the probability of rolling a one.   Now multiply by the number of trials.

$$P(1) = \frac{1}{6}$$

$$\frac{1}{6} \times 90$$
$$= 90 \div 6$$
$$= 15$$

We would expect to roll a one 15 times.

2   A bag contains 7 red and 3 yellow marbles. A marble is drawn at random and then returned. How many red marbles would you expect to draw if you did this

a)   20 times   b)   50 times?

a) $P(\text{red}) = \frac{7}{10}$

$\frac{7}{10} \times 20 = 14$. We would expect to draw 14 red marbles in 20 attempts.

b) $\frac{7}{10} \times 50 = 35$. We would expect to draw 35 red marbles in 50 attempts.

3   The probability of winning a prize in a school raffle is $\frac{3}{20}$. The probability of winning a prize in a beat-the-goalie competition is 12%. In which competition are you more likely to win?

Remember, to compare fractions, decimals and percentages we must convert them all to the same form.

$$\frac{3}{20} = \frac{15}{100}$$

$= 0{\cdot}15$ $12\% = 0{\cdot}12$ Since 0·15 is larger than 0·12, you are more likely to win in the raffle.

Exercise 14C

A calculator may be used throughout this exercise.

1 How many times would you expect to throw heads if you tossed a coin:

   a) 10 times   **b)** 50 times   **c)** 82 times?

2 How many times would you expect to roll an even number on a six-sided die if you tried:

   a) 30 times   **b)** 100 times   **c)** twice?

3 How many times would you expect the top card of a shuffled deck to be a heart if you tried:

   a) 20 times   **b)** 52 times   **c)** 300 times?

4 The probability that you win a game of 'rock, paper, scissors' against a computer making random selections is $\frac{1}{3}$. How many times would you expect:

   a) to beat the computer in 15 attempts

   b) to have to play before achieving 35 wins

   c) **not** to win in 42 games?

5 A bag contains 3 red, 4 blue, 7 green and 6 purple marbles. A marble is removed at random and then replaced. If this is done 60 times, how many times would you expect to draw:

   a) a red marble   **b)** a blue marble   **c)** a green marble   **d)** a yellow marble?

6 A bag contains red, green and blue marbles. The probability of drawing a red one at random is $\frac{1}{5}$ and the probability of drawing a green one is $\frac{1}{6}$.

   a) What is the probability of drawing a blue marble?

   b) If there are 18 red marbles in the bag, how many marbles are there altogether?

7 As part of a statistical experiment, Kate shuffles a deck of cards and looks at the top card. She repeats this process 260 times. How many times would she expect to see:

   a) a face card   **b)** a diamond   **c)** an ace   **d)** a card with a prime number

   e) an even number   **f)** a black 3, 4 or 5   **g)** the 3 of clubs   **h)** a red king or queen?

8 The contents of three marble bags are shown.
From which bag do you have the greatest chance
of drawing:

   a) a red marble

   b) a black marble

   c) a blue marble?

A   B   C

9 A raffle organised by the school charity committee has 12 prizes. They sell 250 tickets.

   a) Calculate the probability of winning a prize with a single ticket.

   Before the draw, the charity committee sell a further 20 tickets and have one additional prize donated.

   b) What is the probability of winning a prize with a single ticket?

   c) Was a ticket more likely to win a prize before or after the additional donation and ticket sales?

10 Till's marble bag contains 45 red and 75 black marbles. Richard's contains 82 red and 158 black marbles. From whose bag are you more likely to draw a red marble?

Check-up 👍

1   Describe in words the probability of each of the following events.

a)   School ends at 3:30 pm

b)   A number rolled on a 6-sided die is less than 10

c)   At least one pupil is off school today

d)   A flipped coin comes up 'heads' 20 times in a row

e)   A person lives to 100 years old

f)   You turn over an ace from a shuffled deck

g)   A random whole number is even

h)   The sun sets in the east

2   A bag contains 9 green balls and 15 yellow balls. What is the probability that a ball chosen at random is:

a)   green   b)   red   c)   yellow?

3   What is the probability that the top card of a shuffled deck is:

a)   a queen   b)   the 3 of clubs   c)   a heart   d)   not an ace

e)   a red 2, 3 or 4   f)   an even numbered card   g)   **not a club**   h)   not an 8 or 9?

4   On this spinner shown, what is the probability of getting:

a)   a 6   b)   an even number

c)   a prime number   d)   a number less than 3?

5   For each of these bags, write down the probability of drawing a red marble.

a)

b)

c)

6   The probability that Jane goes running in the morning is $\frac{5}{7}$. What is the probability that she does **not** go running?

7   A confectioner designs a mystery chocolate box in which all the chocolates look identical. There are 3 caramel, 3 coffee, 4 raspberry, 4 truffle, 5 hazelnut and 5 coconut chocolates.

a)   What is the probability that the first chocolate picked from the box is a truffle?

b)   What is the probability that the first chocolate picked is a caramel or hazelnut?

c)   If the first two chocolates that are picked out are both truffles, what is the probability that the next is raspberry?

d)   How many chocolates must someone consume to be certain of eating a coconut?

8   A box of colouring pencils contains 3 yellow, 4 red, 4 purple, 6 brown, 8 blue, 1 pink and 6 orange pencils. What is the probability that a pencil chosen at random is:

a)   red   b)   brown   c)   pink   d)   grey

e)   blue or orange   f)   pink, purple or yellow   g)   not blue   h)   not orange or yellow?

→

9 The numbers of goals scored by players at a football club are shown in this chart.

   What is the probability that a player picked at random from the squad has scored:

   a) no goals

   b) ten goals

   c) at least seven goals

   d) fewer than three goals?

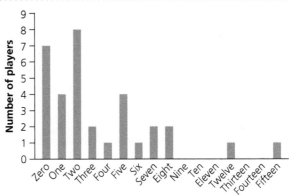

Number of goals

10 A word game uses tiles with letters on them, each worth a specific number of points as shown in the table.

Letter	Points	Number of tiles	Letter	Points	Number of tiles	Letter	Points	Number of tiles
A	1	6	J	4	1	S	1	6
B	3	4	K	3	2	T	1	6
C	2	4	L	1	5	U	1	3
D	2	4	M	2	4	V	4	2
E	1	6	N	1	6	W	4	2
F	2	4	O	1	5	X	5	1
G	2	4	P	1	5	Y	3	2
H	3	4	Q	5	1	Z	5	1
I	1	5	R	1	5	Blank	0	2

There are 100 tiles in total in the game and they are kept in a bag. What is the probability that a tile chosen at random is:

   a) a letter R   b) blank   c) an S or T   d) a vowel

   e) worth 4 points   f) worth less than 2 points   g) the only tile with that letter

   h) a consonant?

11 If you shuffled a deck of cards and turned over the top card 104 times, how many times would you expect to turn over:

   a) the ace of spades   b) an even numbered card   c) a diamond   d) a 7

   e) a red face card   f) a black number card   g) a number greater than 7   h) a red 4?

12 If you roll a 12-sided die 60 times, how many times would you expect to roll:

   a) a 7   b) a 2 or 3   c) an odd number

   d) a 4, 5 or 6   e) a two-digit number   f) a Fibonacci number

   g) a number other than 1   h) a square number?

13 Use a calculator to determine which of the following is most likely:

   ● drawing a red marble from a bag containing 6 red and 9 blue marbles

   ● drawing a black marble from a bag containing 8 black and 14 white marbles

   ● drawing a green marble from a bag containing 13 green and 18 yellow marbles.

Answers

1 Number skills

Exercise 1A

1 a) 0·9   b) 8·6   c) 9·49   d) 55·509
   e) 84·678 f) 66·62 g) 10·123 h) 66·944
2 £3·15   3 92·7 cm
4 a) 0·71   b) 7·93   c) 9·64   d) 26·35
   e) 0·507 f) 1·673 g) 8·849 h) 22·513
   i) 144·954 j) 13·058 k) 14·542 l) 1·989
5 a) 10·9   b) 16·79 c) 75·549
   d) 17·402 e) 48·989 f) 317·913
6 31·93 miles 7 44·754 points

Exercise 1B

1 a) 0·3   b) 5·6   c) 9·23   d) 6·12
   e) 5·251 f) 6·09   g) 2·94   h) 2·58
   i) 0·789 j) 10·886 k) 1·722 l) 0·108
2 38·3 miles
3 a) 2·57   b) 9·22   c) 722·29 d) 29·237
   e) 11·354 f) 4·901 g) 21·495 h) 64·598
   i) 115·566 j) 1·031 k) 15·971 l) 0·382
4 a) 25·1   b) 112·107   c) 2·033
5 £3·55   6 £2·55
7 a) 0·06 seconds   b) Third and fourth
   c) 0·16 seconds   d) 12·01 seconds

Exercise 1C

1 a) 104   b) 312   c) 225   d) 552
   e) 343   f) 982   g) 3195 h) 2124
   i) 8645 j) 12328
2 216 players 3 270 plants 4 1512 ml
5 a) 1·4   b) 11·2   c) 118·5 d) 288·6
   e) 2·04 f) 10·04 g) 136·08 h) 307·08
   i) 8·386 j) 83·62
6 1·8 litres
7 a) 7·07 b) 10·872 c) 53·212 d) 1·194

Exercise 1D

1 a) 12   b) 17   c) 45   d) 139   e) 613
2 250 ml
3 a) 6·9 b) 2·4 c) 24·1 d) 0·51 e) 1·58
4 16·5 cm
5 a) 14·8   b) 17·5   c) 15·5   d) 0·45
   e) 2·05   f) 4·15   g) 3·96   h) 26·35
   i) 1·75   j) 3·942
6 £28.75   7   £1·32
8 a) 250 g   b) 31·25 g   9 0·06 tonnes

Exercise 1E

1 a) 4·8   b) 7·2   c) 27·5   d) 85·4
   e) 523·3 f) 405·8 g) 0·1   h) 2·0
   i) 7·0   j) 10·0
2 a) 5·22   b) 4·39   c) 20·64 d) 204·82
   e) 3147·75 f) 0·48 g) 16·30 h) 0·01
   i) 0·18   j) 3·00
3 a) 1·205 b) 9·081 c) 12·745 d) 305·877
   e) 627·048 f) 0·002 g) 0·075 h) 10·081
   i) 12·010 j) 7·000
4 £1·12   5   6
6 a) £44   b) £14·67   c) £16·34

Exercise 1F

1 a) correct, wrong, wrong, correct, wrong
   b) 26·7 × 4 = 106·8,
      23·193 + 16·82 − 2·58 = 37·433,
      35·267 × 9 = 317·403
2 a) i) Yes   ii) No   iii) No
   b) i) £19   ii) £20·94 iii) £20·15
3 To make sure that you have enough money to
   pay for the items.

Exercise 1G

1 a) 430   b) 71   c) 190·6 d) 9143·45
   e) 8900 f) 15   g) 1687·3 h) 8209·51
   i) 613000 j) 51200 k) 327   l) 4100·9
2 a) 19   b) 6·2   c) 7·48   d) 0·135
   e) 42·69 f) 6·634 g) 0·4761 h) 0·07624
   i) 12·507 j) 0·428 k) 0·01297 l) 0·001025

3 a) 6 b) 10 c) 8 d) 100 e) 2 f) 1000
4 a) 540 b) 2100 c) 32 d) 1272
   e) 252·8 f) 2295·6 g) 1220 h) 8295
   i) 4135·5 j) 6110·1 k) 87500 l) 124070·2
5 £64·50   6 £1900
7 a) 10 b) 5 c) 100 d) 8
   e) 1000 f) 7
8 a) 16 b) 2·6 c) 3·63 d) 0·18
   e) 2·75 f) 4·1 g) 5·27 h) 4·018
   i) 0·825 j) 0·025 k) 0·652 l) 0·0823
9 0·07 m   10 £11·46   11 £4·95
12 a) £43·40   b) 25 days

Exercise 1H

1 a) 1058 b) 1653 c) 4836
   d) 2310 e) 2924
2 1008   3 1800   4 713
5 a) 4212 b) 12862 c) 40678
   d) 320327 e) 106088
6 13 696 miles   7 8000 ml
8 a) 6900 b) 1158

Exercise 1I

1 a) 88·8 b) 6105·4 c) 395·65
   d) 1017·52 e) 2313·6 f) 39078·6
   g) 6245·16 h) 2647·92
2 £344·75   3 £133·30   4 £780·30
5 a) 158274·27   b) 164961·216
   c) 5187   d) 297138·902
6 £20668·40
7 a) £64756·25   b) £13580
   c) £129169·85   d) £207758·30
8 £2412828·60

Exercise 1J

1 a) 5 b) 6 c) 3 d) 8
   e) 3 f) −5 g) −3 h) −4
   i) −6 j) 0
2 4

3 a) −1 b) −4 c) −3 d) −5
   e) −13 f) −4 g) −7 h) −9
   i) −13 j) −18
4 −8 °C
5 a) −30 b) −24 c) −21 d) −42
   e) −32 f) −60 g) −255 h) −93
   i) −52 j) −45
6 −259 °C   7 6 over par

Exercise 1K

1 a) 4 b) 9 c) 22 d) −2
   e) −4 f) −6 g) −4 h) −5
   i) −9 j) −11 k) −7 l) −15
   m) −12 n) −19 o) −27
2 a) 6 b) 4 c) 11 d) 11
   e) 16 f) 29 g) 55 h) 2
   i) 5 j) 2 k) −9 l) −5
   m) −6 n) −10 o) −16
3 a) −9 b) 8 c) −12 d) −4
   e) −7 f) −13 g) −12 h) −15
   i) 5 j) −20 k) −62 l) 31
   m) 97 n) −23 o) −105
4 a) 2 b) −9 c) −4

Exercise 1L

1 a) −10 b) −32 c) −84 d) −150
   e) −20 f) −14 g) −66 h) −160
   i) 24 j) 40 k) 36 l) 78
   m) 90 n) 288 o) 415
2 a) −24 b) −30 c) 72 d) 105
3 a) 102 b) −90 c) 261 d) −945

Exercise 1M

1 a) −2 b) −3 c) −7 d) −12
   e) −3 f) −8 g) −5 h) −30
   i) −26 j) −21 k) −41 l) −132
   m) 4 n) 11 o) 19
2 a) −13·5 b) 21·8 c) −3·6 d) −27·25
3 a) −3 b) −6 c) −5 d) 2
4 a) 12 and −2 b) 32 and −8 c) −4 and −2

Exercise 1N

1. a) 30   b) 18   c) 18   d) 1
  e) 2   f) 2   g) 10   h) 11
  i) 66   j) 20   k) 22   l) 26
2. a) YOLK   b) STARK   c) HEART
  d) Pupils' own answers
3. a) 18   b) 3·1   c) −8   d) −5
  e) 9   f) 8   g) 1   h) 20
4. a) 5   b) 12   c) 8   d) 220
  e) 8   f) −6

Check-up

1. a) 24·31 b) 1·233 c) 38·4 d) 21·528
  e) 273·6 f) 5·8   g) 2·75 h) 1·62
2. a) No   b) £3·65
3. a) 3·14   b) 3·143
4. a) 0·184 litres   b) 0·39 litres
  c) 0·602 litres   d) 0.22 litres
5. a) 139   b) 42361·8 c) 15647·18
  d) 129   e) 6·35   f) 0·1821
  g) 198·1   h) 3643·2   i) 133680
  j) 23·1   k) 9·024   l) 0·0235
6. £57
7. a) £0·76   b) £95
8. 1620 metres   9   £64·17
10. a) approx. 10950 b) 10585 c) £529·25
11. a) −4   b) −12   c) 3   d) 14
  e) −8   f) 1   g) −23 h) −9
12. 42 °C
13. a) −14 b) −20 c) −48 d) −9
  e) 10   f) −4   g) −7   h) 20
14. a) 57   b) 2   c) 12   d) 2
15. a) Thomas b) Mairi c) Fiona
16. a) $2 \times (6 - 1) + 7$ b) $(2 - 6) \times 7 + 1$
  c) $(6 + 2) \times 7 - 1$ d) $1 - (6 + 2) \times 7$

2 Multiples, factors, powers and roots

Exercise 2A

1. a) 7, 14, 21, 28, 35
  b) 3, 6, 9, 12, 15
  c) 9, 18, 27, 36, 45
  d) 8, 16, 24, 32, 40
  e) 12, 24, 36, 48, 60
  f) 15, 30, 45, 60, 75
  g) 40, 80, 120, 160, 200
  h) 18, 36, 54, 72, 90
  i) 26, 52, 78, 104, 130
  j) 104, 208, 312, 416, 520
2. 105   3 2028   4 Yes   5 No
6. a) 2   b) 5   c) both
  d) neither e) both   f) 5
7. It ends in an even digit.
8. It ends in either a 0 or a 5.
9. a) yes   b) yes   c) no   d) yes
  e) no   f) yes   g) yes   h) no
10. a) 18   b) 12   c) 35   d) 24
  e) 20   f) 48   g) 56   h) 120
  i) 42   j) 39   k) 168   l) 143
11. a) 60   b) 70   c) 120   d) 168
12. 36 seconds   13 5805
14. 11 pm on Tuesday   15 1:30 pm
16. 120   17 27 pupils   18 57 treats

Exercise 2B

1. a) 1, 2, 5, 10   b) 1, 3, 5, 15
  c) 1, 2, 3, 4, 6, 8, 12, 24 d) 1, 3, 13, 39
  e) 1, 17
  f) 1, 2, 4, 5, 10, 20, 25, 50, 100
  g) 1, 3, 9, 27, 81
  h) 1, 2, 3, 6, 9, 18, 27, 54
  i) 1, 2, 3, 4, 5, 6, 9, 10, 12, 15, 18, 20, 30, 36, 45, 60, 90, 180

j) 1, 2, 4, 5, 8, 10, 16, 20, 25, 40, 50, 80, 100, 200, 400

k) 1, 3, 5, 7, 15, 21, 35, 105   l) 1

2 Square numbers, e.g. 16, 25, 49

3 a) 7   b) 4   c) 15   d) 8
  e) 27   f) 1   g) 2   h) 13
  i) 35   j) 17   k) 6   l) 7

4 a) 3   b) 7   c) 12

5

1st number	2nd number	Product	Highest common factor	Lowest common multiple
6	15	90	3	30
8	12	96	4	24
20	50	1000	10	100
18	36	648	18	36
22	33	726	11	66

The product of two numbers is the product of their LCM and HCF.

6 28

7 Abundant numbers can be partitioned several ways.

8 15th of November

9 This is true for all pairs of positive integers.

Exercise 2C

1 a) yes   b) no   c) no   d) yes
  e) yes   f) no   g) yes   h) no
  i) no   j) no   k) no   l) no

2

	2	3		5		7			
11		13				17		19	
		23						29	
31						37			
41		43				47			
		53						59	
61						67			
71		73						79	
		83						89	
						97			

3 a) 2   b) 2   c) 8

4 a) 11 and 13, 17 and 19, 29 and 31, 41 and 43, 59 and 61, 71 and 73
  b) e.g. 101 and 103, 107 and 109

5 a) 2 × 5   b) 3 × 7   c) 5 × 11
  d) 2 × 17   e) 7 × 11   f) 11 × 13

6 a) 3 + 5   b) e.g. 3 + 17   c) e.g. 11 + 17
  d) e.g. 3 + 37 e) e.g. 3 + 61 f) e.g. 47 + 53

7 e.g. $3^2 + 4^2 = 25$

8 None: all arrangements are divisible by 3

9 11, 13, 17, 31, 37, 71, 73, 79, 97

10 11, 23, 29, 41, 43, 47, 61, 67, 83, 89

11 a) 64   b) 131

Exercise 2D

1 a) yes   b) no   c) no   d) no   e) yes

Exercise 2E

1 a) 9^2 b) 8^3 c) 5^4 d) 2^4
  e) 3^3 f) 11^4 g) a^4 h) b^3

2 a) $2^4 × 3^2$ b) $5^2 × 7^2 × 13^2$
  c) $2 × 3^2 × 7^3$ d) $11^2 × 17^3$
  e) $23^2 × 29^3$ f) $2^4 × 3^2 × 5 × 7^2$

3 a) 8   b) 36   c) 81   d) 64
  e) 81   f) 1   g) 216   h) 32
  i) 1000000 j) 343   k) 256   l) 8000

4 a) 34   b) 8   c) 1027   d) 37
  e) 28   f) 25   g) 36   h) 31
  i) 91   j) 45   k) 17   l) 48

5 a) 64   b) −27   c) 10000 d) −216
  e) −1   f) −32   g) 400

6 A negative raised to an odd power is negative. A negative raised to an even power is positive.

7 a) 15, no b) 31, yes c) 127, yes d) 255, no

8 No, as multiplying two odd numbers gives an odd answer.

9 a) 35831808 b) 262144   c) 1889568
  d) 961   e) 4913   f) 390625
  g) 1024   h) 4477456   i) 1048576
  j) 117649   k) 2744   l) 59049

10 a) 1269   b) 769   c) 21455   d) 1707
   e) 99   f) 5400   g) 405   h) 729

Exercise 2F

1 a) ±6   b) ±2   c) ±7   d) ±10   e) ±8
   f) ±11   g) ±20   h) ±12   i) ±15   j) ±14
2 a) 2   b) 1   c) 10   d) −3   e) 5

Exercise 2G

1 a) $2^2 \times 5$ b) $3^2 \times 5$ c) $2 \times 3 \times 5$
   d) $3^2 \times 7$ e) 2×3^3 f) $2^2 \times 3 \times 7$
   g) 3×5^3 h) $2^3 \times 5$ i) $2^4 \times 3^2$
   j) $2^2 \times 3 \times 11$ k) $2^2 \times 5^2$ l) $3 \times 7 \times 11$
   m) $2^2 \times 3 \times 5^2$ n) $2^2 \times 3 \times 7^2$ o) $2^6 \times 3$
   p) $2^2 \times 5^2 \times 7$
2 a) $80 = 2^4 \times 5, 207 = 3^2 \times 23$, yes
   b) $544 = 2^5 \times 17, 675 = 3^3 \times 5^2$, yes
   c) $744 = 2^3 \times 3 \times 31, 945 = 3^3 \times 5 \times 7$, no
   d) $183 = 3 \times 61, 3040 = 2^5 \times 5 \times 19$, yes

Exercise 2H

1 a) HCF = 6, LCM = 720
   b) HCF = 2, LCM = 570
   c) HCF = 12, LCM = 120
   d) HCF = 3, LCM = 8100
   e) HCF = 9, LCM = 2916
   f) HCF = 21, LCM = 4725
   g) HCF = 4, LCM = 24200
   h) HCF = 18, LCM = 8100
2 24   3 12   4 18

Check-up

1 a) 4, 8, 12, 16, 20   b) 5, 10, 15, 20, 25
   c) 11, 22, 33, 44, 55   d) 13, 26, 39, 52, 65
   e) 25, 50, 75, 100, 125
   f) 60, 120, 180, 240, 300
   g) 120, 240, 360, 480, 600
   h) 115, 230, 345, 460, 575
2 a) 42   b) 60   c) 140   d) 189
3 180   4 126
5 a) 60 minutes   b) 5, 4 and 3 laps

6 a) 274th beat   b) 54
7 Yes, digits sum to a multiple of 3.
8 No, digits don't sum to a multiple of 3.
9 No, doesn't end in 0 or 5.
10 a) 1, 2, 7, 14   b) 1, 2, 4, 8, 16
   c) 1, 5, 17, 85
   d) 1, 2, 4, 5, 10, 11, 20, 22, 44, 55, 110, 220
11 a) 6   b) 28   c) 36   d) 75
12 8   13 21
14 a) yes   b) no: 1, 3, 17, 51
   c) no: 1, 2, 13, 26   d) no: 1, 3, 31, 93
15 a) yes   b) no, 8   c) no, 9   d) no, 11
16 8624
17 a) 6^4 b) 8^3 c) 13^2 d) 2^7
   e) 4^5 f) 53^4 g) 3^3 h) a^5
18 a) 81   b) 27   c) 64   d) 625
   e) 10000   f) 343   g) 16   h) −1
19 12   20 4
21 a) 2×5^2 b) $3^3 \times 5$ c) 3^4
   d) 2×7^2 e) $2 \times 3^3 \times 7$ f) $2 \times 3^2 \times 11$
   g) $2^3 \times 3 \times 11$ h) $2^2 \times 5 \times 7^2$
22 a) HCF = 5, LCM = 650
   b) HCF = 3, LCM = 198
   c) HCF = 12, LCM = 72
   d) HCF = 9, LCM = 324
   e) HCF = 10, LCM = 910
   f) HCF = 1, LCM = 690
   g) HCF = 13, LCM = 195
   h) HCF = 25, LCM = 600
23 36
24 a) 158   b) 244   c) 123   d) 192

3 Fractions, decimals and percentages

Exercise 3A

1 a) 5   b) 8   c) 12   d) 3
2 a) 3   b) $\frac{2}{3}$ c) $\frac{1}{20}$ d) $\frac{1}{12}$
   e) $\frac{10}{9}$ f) $\frac{8}{7}$ g) $\frac{1}{3}$

3 $\dfrac{7}{30}$ 4 $\dfrac{22}{25}$ 5 $\dfrac{1}{5}$ 6 $\dfrac{5}{8}$ 7 $\dfrac{9}{10}$

Exercise 3B

1 a) $\dfrac{1}{10}$ b) $\dfrac{1}{5}$ c) $\dfrac{4}{5}$ d) $\dfrac{14}{25}$

   e) $\dfrac{3}{4}$ f) $\dfrac{11}{100}$ g) $\dfrac{99}{100}$ h) $\dfrac{3}{1000}$

   i) $\dfrac{1}{200}$ j) $\dfrac{2}{125}$ k) $\dfrac{6}{125}$ l) $\dfrac{51}{500}$

2 $0.2 + 0.1 = \dfrac{3}{10}$, $0.8 - 0.32 = \dfrac{12}{25}$, $0.96 \div 8 = \dfrac{3}{25}$,

   $0.24 \times 3 = \dfrac{18}{25}$, $0.02 \times 3 = \dfrac{3}{50}$,

   $0.209 + 0.159 = \dfrac{46}{125}$, $0.9 + 0.02 = \dfrac{23}{25}$,

   $1 - 0.275 = \dfrac{29}{40}$

3 a) 50%   b) 90%   c) 2%   d) 12%
   e) 64%   f) 0.4%   g) 4.5%   h) 66%
   i) 8%   j) 80.6%   k) 20.5%   l) 95.4%

Exercise 3C

1 a) $\dfrac{9}{100}$ b) $\dfrac{19}{100}$ c) $\dfrac{51}{100}$ d) $\dfrac{67}{100}$

   e) $\dfrac{91}{100}$ f) $\dfrac{17}{100}$ g) $\dfrac{37}{100}$ h) $\dfrac{2}{25}$

   i) $\dfrac{23}{50}$ j) $\dfrac{41}{50}$ k) $\dfrac{19}{20}$ l) $\dfrac{17}{25}$

   m) $\dfrac{9}{25}$ n) $\dfrac{27}{50}$

2 a) 0.06   b) 0.3   c) 0.72   d) 0.59
   e) 0.83   f) 0.92   g) 0.16   h) 0.105
   i) 0.762   j) 0.9438   k) 0.4431   l) 0.0716
   m) 0.008   n) 0.0005

3 $\dfrac{17}{50}$ 4 $\dfrac{19}{50}$ 5 $\dfrac{3}{20}$

6 a)

0.4	71%	14%	0.3	19%	24%	0.57
0.37	12%	62%	59%	0.46	61%	2%
33%	9%	0.38	0.17	16%	0.34	91%
0.02	3%	0.67	10%	20%	0.15	0.77
54%	0.87	13%	0.65	0.14	53%	12%
0.83	67%	0.55	19%	34%	8%	0.01

   b) 16%   c) 91% + 0.77 + 12% + 0.01

Exercise 3D

1 a) 0.9   b) 0.6   c) 3.5   d) 0.625
   e) 2.2   f) 0.45   g) 0.16

2 a) 0.667   b) 0.556   c) 0.857   d) 0.889
   e) 1.429   f) 4.667   g) 0.633

3 a) 58%   b) 70%   c) 55%   d) 80%
   e) 12.5%   f) 32.5%   g) 61.25%

4 a) $\dfrac{51}{100}$, 0.52, $\dfrac{14}{25}$, 0.57, 58%

   b) 83%, $\dfrac{21}{25}$, $\dfrac{17}{20}$, 87%, 0.88

   c) 41%, 0.42, $\dfrac{9}{20}$, $\dfrac{23}{50}$, 49%

   d) $\dfrac{1}{200}$, 0.6%, 5%, 0.052, $\dfrac{29}{500}$

Exercise 3E

1 a) 6   b) 14   c) 24   d) 35
   e) £57   f) 113 kg   g) 30 m   h) 12 cm

2 a) 10   b) 9   c) 48   d) 77
   e) 115   f) 162   g) 750   h) 2.5
   i) 6.4   j) 7.5   k) 5.4   l) 5.6

3 a) £3   b) 7 kg   c) 1.5 ml   d) 2.6 miles

4 a) 77 miles   b) £76   c) £37.68   d) 77 m

5 £0.80   6 90 minutes   7 88 pupils

8 480 g   9 £31.50   10 £172500

11 a) 1500   b) 300

12 a) 7 bags   b) £159   c) 5·25 square metres

13 £450   14 21 pets   15 360 spectators

16 a) £6·25   b) £12·50   c) £11·25

Exercise 3F

1 a) 4   b) 24   c) 40   d) 60
  e) 88   f) 400   g) 800

2 a) 6   b) 24   c) 114   d) 300
  e) 432   f) 612   g) 3000

3 a) $\dfrac{9}{2}$ b) $\dfrac{13}{6}$ c) $\dfrac{19}{5}$ d) $\dfrac{9}{7}$

  e) $\dfrac{35}{4}$ f) $\dfrac{71}{11}$ g) $\dfrac{61}{8}$ h) $\dfrac{115}{12}$

  i) $\dfrac{52}{5}$ j) $\dfrac{59}{10}$ k) $\dfrac{65}{4}$ l) $\dfrac{142}{7}$

  m) $\dfrac{106}{3}$ n) $\dfrac{1312}{13}$

4 a) correct   b) correct   c) wrong
  d) correct   e) correct   f) 80%

Exercise 3G

1 a) 6   b) 4   c) 12   d) 7
  e) 16   f) 24   g) 80

2 a) $8\dfrac{1}{2}$ b) $3\dfrac{2}{3}$ c) $4\dfrac{4}{5}$ d) $4\dfrac{3}{4}$

  e) $3\dfrac{3}{10}$ f) $6\dfrac{1}{8}$ g) $12\dfrac{5}{6}$ h) $5\dfrac{1}{2}$

  i) $4\dfrac{1}{3}$ j) $6\dfrac{1}{3}$ k) $5\dfrac{3}{4}$ l) $7\dfrac{3}{5}$

  m) $30\dfrac{1}{2}$ n) $4\dfrac{9}{10}$

3 $6\dfrac{1}{5}, 6\dfrac{1}{4}, 6\dfrac{2}{7}, 6\dfrac{1}{2}, 6\dfrac{4}{7}, 6\dfrac{2}{3}, 6\dfrac{4}{5}$.

4 a) FDAEC – B is the odd one out.
  b) CBFDA
  c) No, every card can be placed. They can always be linked using a repeated block of FDAECB.

Exercise 3H

1 a) $\dfrac{4}{7}$ b) $\dfrac{3}{5}$ c) $\dfrac{2}{3}$ d) $\dfrac{3}{4}$ e) $\dfrac{4}{5}$

---

f) 1   g) $\dfrac{5}{9}$ h) $\dfrac{1}{5}$ i) $\dfrac{1}{2}$ j) $\dfrac{2}{3}$

k) $\dfrac{1}{3}$ l) $\dfrac{2}{5}$

2 a) $1\dfrac{2}{5}$ b) $1\dfrac{5}{11}$ c) $1\dfrac{4}{15}$ d) $2\dfrac{5}{9}$

  e) $2\dfrac{3}{4}$ f) $2\dfrac{2}{5}$

3 a) $\dfrac{11}{13}$ b) $1\dfrac{9}{11}$ c) 2   d) $1\dfrac{1}{3}$

4 a) $\dfrac{2}{3}$ b) $\dfrac{1}{3}$

5 a) $\dfrac{6}{7}$ b) $\dfrac{3}{7}$ c) 120

6 a) $\dfrac{1}{2}$ b) $\dfrac{1}{5}$ c) $\dfrac{3}{4}$

7 a) $\dfrac{5}{11}$ and $\dfrac{4}{11}$ b) $\dfrac{4}{5}$ and $\dfrac{1}{5}$

8 a)

$\dfrac{3}{13}$	$\dfrac{7}{13}$	$\dfrac{2}{13}$
$\dfrac{3}{13}$	$\dfrac{4}{13}$	$\dfrac{5}{13}$
$\dfrac{6}{13}$	$\dfrac{1}{13}$	$\dfrac{5}{13}$

b)

$\dfrac{2}{19}$	$\dfrac{7}{19}$	$\dfrac{6}{19}$
$\dfrac{9}{19}$	$\dfrac{5}{19}$	$\dfrac{1}{19}$
$\dfrac{4}{19}$	$\dfrac{3}{19}$	$\dfrac{8}{19}$

c)

$\dfrac{8}{21}$	$\dfrac{1}{7}$	$\dfrac{4}{21}$
$\dfrac{1}{21}$	$\dfrac{5}{21}$	$\dfrac{3}{7}$
$\dfrac{2}{7}$	$\dfrac{1}{3}$	$\dfrac{2}{21}$

d)

$\frac{1}{3}$	$\frac{8}{9}$	$\frac{1}{9}$
$\frac{2}{9}$	$\frac{4}{9}$	$\frac{2}{3}$
$\frac{7}{9}$	0	$\frac{5}{9}$

Exercise 3I

1 a) 12   b) 42   c) 18   d) 12   e) 210

2 a) 10   b) 6   c) 30   d) 18   e) 42

3 a) 3   b) 20   c) 54   d) 60
   e) 44   f) 80

4 a) $\frac{3}{4}$ b) $\frac{3}{4}$ c) $\frac{2}{3}$ d) $1\frac{1}{2}$
   e) $1\frac{1}{6}$ f) $\frac{25}{36}$ g) $1\frac{11}{42}$ h) $1\frac{23}{60}$
   i) $1\frac{31}{40}$ j) $\frac{47}{80}$ k) $\frac{77}{78}$ l) $1\frac{49}{60}$

5 a) $\frac{5}{8}$ b) $\frac{3}{10}$ c) $\frac{5}{9}$ d) $\frac{1}{8}$
   e) $\frac{4}{21}$ f) $\frac{33}{140}$

6 a) $1\frac{2}{3}$ b) $\frac{11}{30}$ c) $\frac{9}{40}$ d) $\frac{11}{18}$
   e) $\frac{9}{40}$ f) $\frac{11}{210}$ g) $\frac{43}{60}$ h) 0

7 a) $\frac{13}{20}$ b) $\frac{5}{12}$ c) $\frac{11}{60}$

8 a) $\frac{46}{63}$ b) $\frac{7}{36}$ c) $\frac{25}{36}$

9 a) $\frac{5}{12}$ and $\frac{1}{3}$ b) $\frac{2}{3}$ and $\frac{7}{15}$ c) $\frac{9}{10}$ and $\frac{2}{5}$

10 a) $\frac{4}{9}, \frac{4}{9}, \frac{7}{9}$ or $\frac{4}{9}, \frac{11}{18}, \frac{11}{18}$
    b) $\frac{2}{3}, \frac{1}{5}, \frac{2}{5}$ or $\frac{2}{3}, \frac{1}{3}, \frac{4}{15}$

Exercise 3J

1 a) $\frac{1}{6}$ b) $\frac{7}{48}$ c) $\frac{6}{77}$ d) $1\frac{19}{26}$

e) $1\frac{11}{54}$ f) $1\frac{19}{65}$

2 a) $\frac{2}{5}$ b) $\frac{1}{21}$ c) $\frac{2}{3}$ d) $\frac{3}{10}$
   e) $\frac{8}{21}$ f) $5\frac{5}{6}$

3 a) $\frac{3}{5}$ b) $\frac{1}{6}$ c) $\frac{1}{4}$ d) $\frac{1}{12}$
   e) $1\frac{1}{6}$ f) $6\frac{2}{3}$

4 a) $2\frac{2}{3}$ b) $3\frac{3}{5}$ c) 6   d) $2\frac{1}{2}$
   e) $12\frac{2}{3}$ f) $4\frac{1}{2}$

5 $\frac{8}{15}$ 6 $\frac{1}{12}$

7 a) 4   b) $1\frac{1}{2}$ c) $3\frac{1}{2}$ d) $\frac{8}{9}$

Exercise 3K

1 a) $\frac{3}{4}$ b) $\frac{6}{7}$ c) $\frac{5}{18}$ d) $\frac{25}{36}$
   e) $2\frac{4}{7}$ f) $3\frac{3}{10}$

2 a) $\frac{4}{7}$ b) $\frac{2}{9}$ c) $\frac{9}{14}$ d) $\frac{7}{15}$
   e) $1\frac{1}{2}$ f) $2\frac{2}{7}$

3 a) $\frac{2}{3}$ b) $\frac{7}{8}$ c) $\frac{2}{5}$ d) $1\frac{2}{3}$
   e) $1\frac{1}{2}$ f) $1\frac{2}{3}$

4 a) 6   b) $3\frac{1}{3}$ c) $13\frac{1}{2}$ d) $\frac{1}{21}$
   e) $\frac{1}{16}$ f) $\frac{3}{10}$

5 $8\frac{3}{4}$ cm   6 800   7 67 full lengths

8 a) 80   b) $\frac{5}{7}$ c) $2\frac{1}{18}$ d) $11\frac{1}{2}$

Exercise 3L

1 a) 10·75 b) 13·32 c) 469·44 d) 464·8
2 a) 12·8   b) 117·75   c) 126·16
   d) 103·024 e) 637·67   f) 0·684
   g) 2·2632 h) 0·168
4 £4·66   5 £86·63
6 a) 0·8866 seconds   b) 67·3 seconds
7 a) £2572·50 b) £2567·75 c) £4·75
8 a) £10300·90   b) £2199·10

Exercise 3M

1 a) 8   b) 27   c) 20   d) 40
   e) 1·8   f) 4·2   g) 5·6   h) 0·45
2 a) 80   b) 630   c) 180   d) 160
   e) 45   f) 12   g) 48   h) 3·6
3 £32   4 260 g
5 a) 3   b) 8   c) 6, 9   d) 10%, 1%
6 a) 120   b) 296   c) 312   d) 260
   e) 567   f) 147   g) 192   h) 460
7 4440
8 a) 24·6   b) 23·4   c) 29·2   d) 11·6
   e) 27·3   f) 8·4   g) 0·24   h) 2·24
   i) 4·26   j) 3·15   k) 2·82   l) 6·16
9 279   10 986
11 No, it has 2·4 g more saturated fat than recommended in one whole day.

Check-up

1 a) $\dfrac{2}{5}$ b) $\dfrac{3}{4}$ c) $\dfrac{1}{3}$ d) $\dfrac{3}{4}$

   e) $2\dfrac{1}{4}$ f) $5\dfrac{1}{2}$ g) $1\dfrac{11}{20}$

2 $\dfrac{13}{50}$

3 a) $\dfrac{4}{5}$ b) $\dfrac{13}{50}$ c) $\dfrac{43}{50}$ d) $\dfrac{77}{100}$

   e) $\dfrac{21}{200}$ f) $\dfrac{13}{250}$ g) $\dfrac{3}{500}$

4 a) 80% b) 26% c) 86% d) 77%
   e) 10·5% f) 5·2% g) 0·6%

5 a) $\dfrac{17}{100}$ b) $\dfrac{9}{20}$ c) $\dfrac{51}{100}$ d) $\dfrac{3}{50}$

   e) $\dfrac{47}{50}$ f) $\dfrac{18}{25}$ g) $\dfrac{31}{50}$

6 a) 0·17 b) 0·45 c) 0·51 d) 0·06
   e) 0·94 f) 0·72 g) 0·62

7 a) 0·7 b) 0·4 c) 0·6 d) 0·625
   e) 0·45 f) 0·14 g) 0·5125

8 a) 70% b) 40% c) 60% d) 62·5%
   e) 45% f) 14% g) 51·25%

9 a) 0·3, $\dfrac{31}{100}$, 0·32, 34%, $\dfrac{7}{20}$

   b) 0·73, $\dfrac{37}{50}$, $\dfrac{3}{4}$, $\dfrac{19}{25}$, 77%

10 a) 120°   b) 36 m   c) 260 ml
    d) 436 kg   e) 100 miles   f) £558·25

11 a) 275   b) 242

12 a) $\dfrac{15}{2}$ b) $\dfrac{17}{3}$ c) $\dfrac{34}{9}$ d) $\dfrac{37}{8}$

    e) $\dfrac{69}{10}$ f) $\dfrac{51}{4}$

13 a) $4\dfrac{2}{3}$ b) $3\dfrac{3}{5}$ c) $6\dfrac{5}{7}$ d) $5\dfrac{1}{3}$

    e) $6\dfrac{1}{2}$ f) $12\dfrac{2}{3}$

14 a) $\dfrac{1}{2}$ b) $\dfrac{2}{3}$ c) 1 d) $\dfrac{3}{4}$

    e) $\dfrac{2}{5}$ f) $\dfrac{7}{12}$

15 a) $\dfrac{7}{12}$ b) $1\dfrac{5}{8}$ c) $1\dfrac{1}{10}$ d) $1\dfrac{5}{12}$

    e) $1\dfrac{5}{24}$ f) $1\dfrac{31}{45}$ g) $\dfrac{1}{2}$ h) $\dfrac{1}{6}$

    i) $\dfrac{7}{18}$ j) $\dfrac{5}{24}$ k) $\dfrac{2}{35}$ l) $\dfrac{1}{30}$

16 a) $\dfrac{8}{15}$

    b) His son was wrong as this is more than one half.

17 a) $\dfrac{1}{24}$ b) $\dfrac{7}{24}$

18 a) $\dfrac{5}{14}$ b) $\dfrac{1}{7}$ c) $\dfrac{5}{14}$

19 a) $\dfrac{1}{35}$ b) $\dfrac{3}{7}$ c) $\dfrac{5}{12}$ d) $\dfrac{3}{4}$

    e) $\dfrac{5}{21}$ f) $\dfrac{5}{7}$

20 a) $1\dfrac{17}{18}$ b) $\dfrac{7}{8}$ c) $1\dfrac{3}{8}$ d) $\dfrac{3}{8}$

    e) $1\dfrac{1}{5}$ f) $\dfrac{2}{3}$

21 $\dfrac{1}{16}$ 22 13 glasses   23 £15·06

24 a) 18 g   b) 10 mm   c) 49 km
    d) £226·80   e) £283·92   f) £22·68

25 £1090·60

26 a) 40%   b) £22·50

4 Ratio and proportion

Exercise 4A

1 a) 2:3   b) 2:1   c) 4:3   d) 2:5
   e) 1:4   f) 4:9   g) 1:2   h) 9:8

2 a) 2:5   b) 9:4   c) 3:4   d) 7:8
   e) 9:2   f) 1:2   g) 7:2   h) 5:9

3 a) sheep:cattle   b) adults:pups
      10:1   2:1
   c) adults:chicks   d) pigeons:crows
      1:5   35:16

4 Summer Hit   Fruit Treat
   pineapple:mango   apple:orange
      4:3   4:1

5 a) 2:3:5   b) 5:6:8   c) 2:1:10
   d) 11:5:14   e) 8:3:15   f) 1:3:8
   g) 15:1:5   h) 32:43:26

6 The recipes are the same, butter, flour and sugar in the ratio 2:5:1.

7 a) 20   b) dogs:total
          2:5

8 1:11

9 a) children:teenager:adults
     1:15:8
   b) adults:children + teenagers
     1:2
   c) teenager:adults + children
     5:3

10   birth:adult
  grey seal   3:50
  capercaillie   1:23
  red deer   1:19
  pine martin   1:14
  terrier   2:25

11 a) 3:4   b) 1:5   c) 3:2   d) 4:15
   e) 3:5   f) 8:1   g) 2:25   h) 3:400
   i) 1:7   j) 7:2   k) 1:24   l) 1:252

12 5:6

13 Unhappy, 3:50 ratio of adverts to shows which is more than 1:50.

Exercise 4B

1 a) 18   b) 49   c) 72   d) 108
   e) 120   f) 225   g) 280   h) 77

2 a) 12 adults   b) 70 white
   c) 126 flats   d) 1200 guests

3 45 pupils

4 11 teachers   5 160 ml pear puree

6 a) 225 ml water   b) 90 ml juice

7 a) double each side   b) 660 ml treatment
   c) 90 ml dye

8 a) 2000 g sand   b) 1218 g cement

9 a) 57 adults, 21 teenagers b) 93 people

10 84 model atoms   11 £5400

12 608 transactions   13 130 hours

14 4·5 kg oats costs £9, 0·5 kg fruit costs £4, Total = £13

15 a) No, 2l milk requires 500 ml puree

    b) 1500 ml milkshake (300 ml puree, 1200 ml milk)

16 a) No, 140 kg zinc needed

    b) 228 kg copper and 133 kg zinc, Maximum = 361 kg brass

17 150 ml : 225 ml   125 ml : 500 ml, new 275 ml, 725 ml   11 : 29

Exercise 4C

1   a) 20, 30   b) 12, 42   c) 45, 18

    d) 320, 400   e) 210, 60   f) 66, 55

    g) 182, 78   h) 900, 180

2   £14 : £35

3   120 g, 160 g

4   219 days training, 146 days rest

5   63 kg, 45 kg

6   130 days working, 50 days off

7   £20250 reinvested

8   a) 22 m, 44 m, 55 m   b) 135 g, 45 g, 180 g

    c) £10·80, £1·20, £12   d) 4 kg, 2·4 kg, 1·6 kg

    e) 140 mm, 40 mm, 100 mm

    f) 12 hours : 9 hours : 3 hours

    g) 40 min, 15 min, 5 min

    h) 48 min, 36 min, 96 min

9   54 m², 24 m², 12 m²

10 1075 g, 430 g, 215 g

11 £102600 Scotland, £5400 Wales, £108000 England

12 a) 1 litre   b) 3 litres

    c) 1·04 litres   d) 0·84 litres

13 30 more pupils   14 840 bricks

15 520 more spaces

16 a) 36 ml coffee, 144 ml water

    b) 75 ml coffee, 150 ml cold milk, 150 ml cold water, 75 ml foam

    c) 330 ml

d) A regular Cappuccino has more foam with 132 ml, large Frappe 75 ml.

e) Cappuccino contains the most with 132 ml of coffee, Latte 110 ml, Americano 66 ml, Frappe 55 ml.

17 $\frac{3}{4}$ 18 $\frac{2}{9}$ 19 $\frac{2}{5}$

20 a) $\frac{3}{4}$ b) $12\frac{1}{2}\%$

21 a) $\frac{1}{5}$ b) 40%

    c) Cappuccino $\frac{2}{5} = \frac{12}{30}$ = 40% coffee,

        Latte $\frac{1}{3} = \frac{10}{30} = 33\frac{1}{3}\%$ coffee,

        Americano $\frac{1}{5} = \frac{6}{30}$ = 20% coffee,

        Frappe $\frac{1}{6} = \frac{5}{30} = 16\frac{2}{3}\%$

22 7 : 11 : 13 56 stars, 88 stars, 104 stars
    5 : 3 : 23 40 stars, 24 stars, 184 stars

23 £500, £550, £605

Exercise 4D

1   a) 1, 2 8, 16   b) 1, £7 12, £84

    c) 5, 3 40, 24   d) 300 g

    e) 6, 10 30, 50   f) £30

2   320 g   3 350 trees   4 £585

5   160 bins   6 30 minutes   7 40 people

8   900 leaflets   9 495 m³

10 a) Yes, £2·84 each

    b) No, 12 should cost £306

    c) No, it should cool a further 30 °C

    d) Yes 1·9 kg per penguin

    e) No, extra day

11 40 people   12 36 minutes

13 a) £37·50   b) 15 hours

    c) 7·50, 15, 20·50, 30

d) 15, 30, 45, 60

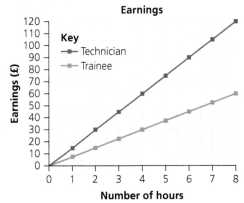

Earnings

Key
- ■ Technician
- ■ Trainee

e) The lines grow further apart; the gap between their income grows with time.

14 a) 260 canapes, 130 favours   b) £1690

15 1350 m²

Exercise 4E

1   10 days   2   6 hours   3   24 days

4   £350   5   30 minutes 6   6 months

7   2·6 m²   8   4 minutes 30 seconds

Check-up

1   a) 3:4   b) 4:15   c) 1:3

    d) 7:20:4   e) 7:2:10   f) 7:10:2

2   3:2   3   2:5   4   8:3

5   1:4   6   1:8   7   150 guests

8   a) 48 g   b) 110 g

9   a) 42 white roses   b) 56 red roses

10   60 ml mango, 150 ml apple

11   a) £24000 stock, £40000 savings

    b) £128000

12   a) 40 m, 16 m   b) 750 g, 150 g

    c) £140, £210   d) £22·50, £13·50, £9

13   1750 g apricots, 1050 g strawberries

14   320 tonnes plastic, 160 tonnes paper, 80 tonnes tin

15   55 more patients   16   £40800, £7200, £24000

17   List 1 C, List 2 B, List 3, A

18   $\frac{1}{8}$ 19 $\frac{1}{8}$

20   a) 6:5:2   b) 2800

    c) 39 cardio, 13 martial, 26 strength

21   1000 litres   22   440 miles

23   32 circuit boards   24   28 bags

25   3 hours   26   6 days

27   a) £990   b) $\frac{1}{5}$ c) £35   d) 5 minutes

5 Time

Exercise 5A

1   a) 4 hours 15 minutes b) 3 hours 35 minutes

    c) 8 hours 5 minutes   d) 21 hours 49 minutes

2   6 hours and 50 minutes

3   a) 6 hours 12 minutes b) 4 hours 48 minutes

    c) 7 hours 4 minutes   d) 15 hours 25 minutes

4   a) 2 hours 30 minutes b) 2 kg

5   a) 10 days and 3 hours

    b) 19 days 22 hours and 15 minutes

6   a)

Location	Bus 1	Bus 2	Bus 3	Bus 4
Dunoon	06:27	07:30	07:50	09:00
Sandbank	06:35	07:38	07:58	09:08
Tighnabruaich	07:03	08:06	08:26	09:36
Kames	07:21	08:24	08:44	09:54
Millhouse	07:24	08:27	08:47	09:57
Portavadie	07:32	08:35	08:55	10:05

    b) 49 mins   c) Dunoon and Tighnabruaich

    d) 08:06

7   a) 6 hours 11 minutes   b) 50p   c) 01:17

8   4 days 22 hours and 30 minutes

9   3 days 8 hours and 44 minutes

10   10 days 22 hours and 49 minutes

11   a) 15th July 5 pm   b) 16th July 2 am

Exercise 5B

1   a) 20 mph   b) 13 m/s   c) 250 km/h

    d) 72 mph   e) 475 m/s   f) 3313 km/h

    g) 24 m/s

2   13 m/s   3   40·5 mph   4   1·5 m/s

5   3 km/h

6   a) 0·05 miles per minute   b) 3 mph

Exercise 5C

1 a) 180 miles   b) 100 m
   c) 1200 km   d) 280 miles
   e) 920 m   f) 2860 km
   g) 138000 km
2 60 m   3 1650 miles
4 126 miles   5 3·36 km
6 a) 40 miles   b) 66 km
   c) Thomas by 2000 m
7 86·4 m

Exercise 5D

1 a) 25 hours   b) 13 seconds   c) 84 hours
   d) 130 seconds   e) 1254 hours
   f) 9 hours and 15 minutes
   g) 83 hours and 30 minutes
2 7 seconds   3 30 minutes 4 5:05 pm
5 3 minutes and 45 seconds

Exercise 5E

1 a) 10 m/s   b) 42 miles   c) 4 hours
   d) 91 km/h   e) 112 m   f) 294 km/h
   g) 2 hours and 30 minutes
2 Yes, as 63 mph is less than 70 mph   3 15 miles
4 3 hours and 15 minutes   5 5·145 km
6 92 km/h   7 160 m
8 a) 4:03 pm   b) 10 metres per minute

Exercise 5F

1 a) 50 mph   b) 1 hour
   c) 25 mph   d) quickly
2 a) 12 mph   b) 2 hours and 30 minutes
   c) 2 mph
3 a) delivery 1 = 60 mph, delivery 2 = 20 mph,
      delivery 3 = 70 mph, delivery 4 = 40 mph
   b) 10:45   c) 30 minutes
   d) delivery 3   e) 18:45   f) 1 hour
4 a) 1 hour and 20 minutes   b) 07:50
   c) 125 metres per minute
   d) 75 metres per minute

5

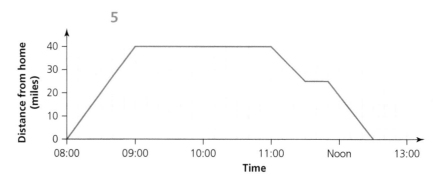

6 Pupils' own answers

Check-up

1 a) 8 hours and 17 minutes
   b) 3 hours and 11 minutes
   c) 13 hours and 39 minutes
   d) 14 hours and 46 minutes
2 1 hour and 42 minutes
3 3 days and 6 hours
4 7 days, 5 hours and 15 minutes
5 a) 60 m/s   b) 500 km/h   c) 52 mph
   d) 16 mm per second
6 a) 180 m   b) 312 miles
   c) 9960 km   d) 3000 cm
7 a) 50 seconds   b) 14 hours
   c) 18 seconds   d) 95 hours
8 175 miles   9 245 km/h
10 1·5 seconds   11 16 mph
12 a) 50 mph   b) 1 hour   c) 60 mph
    d) the second part
13 a) 1 hour and 30 minutes   b) 25 mph
    c) 58 mph   d) 90 miles
14 12 mph   15 330 m   16 5 m/s
17 1 hour and 15 minutes
18 18000000 km
19 10 km/h
20 a) 4·8 km
    b) 5 seconds
21 a) 140 m/min
    b) 1 hour 30 minutes

6 Measurement

Exercise 6A

1. a) 4 cm   b) 5·9 cm   c) 65 cm
   d) 0·8 cm   e) 2·5 cm   f) 15 cm
2. a) 2·4 m   b) 1·26 m   c) 5·2 m
   d) 0·03 m   e) 0·86 m   f) 40 m
3. a) 3 km   b) 2·5 km   c) 0·75 km
   d) 0·08 km   e) 0·007 km   f) 13·7 km
4. a) 6000 m   b) 8400 m   c) 12000 m
   d) 900 m   e) 4071 m   f) 21500 m
5. a) 800 cm   b) 310 cm   c) 60 cm
   d) 1502 cm   e) 18000 cm   f) 2 cm
6. a) 50 mm   b) 42 mm   c) 5 mm
   d) 240 mm   e) 542 mm   f) 1790 mm
7. a) 3·48 m   b) 0·143 km   c) 903·4 m
   d) 7046000 mm   e) 0·075 km
   f) 300000 cm g) 8000 mm   h) 0·001 m
8. a) 2000 g b) 50000 g c) 4071 g d) 40 g
9. a) 8·3 kg   b) 1·27 kg
   c) 0·65 kg   d) 0·054 kg
10. a) 4 l   b) 0·92 l   c) 0·051 l   d) 0·007 l
11. a) 7500 ml   b) 300 ml   c) 7920 ml
    d) 62800 ml
12. a) 4·5 l   b) 7·8 l   c) 13·8 l
    d) 0·304 l
13. a) 50000 cm^2 b) 29000 cm^2
    c) 600000 cm^2 d) 2500 cm^2
14. a) 900 mm^2 b) 6500 mm^2 c) 80 mm^2
15. a) 4000000 cm^3 b) 750000 cm^3
    c) 58000000 cm^3

Exercise 6B

1. a) 45 cm^2 b) 91 cm^2 c) 81 cm^2 d) 900 m^2
2. a) 17·4 m^2 b) 82·2 cm^2
   c) 13 m^2, 130000 cm^2
   d) 12·5 m^2, 125000 cm^2
3. 44100 mm^2
4. a) 30 m^2 120 m^2 210 m^2 b) 1 h 50 min

5. a) 49 cm^2 b) 1600 mm^2
   c) 121 cm^2 d) 225 m^2
6. a) a = 3 cm   b) b = 12 m
   c) c = 6 mm   d) d = 30 cm
7. a) 12800 cm^2 b) 16000 cm^2
   c) 9600 cm^2 d) 38400 cm^2
   e) $33\frac{1}{3}$%   f) $\frac{1}{4}$
   g) $1 - \left(\frac{1}{4} + \frac{1}{3}\right)$ or simplify $\frac{16000}{38400}$

Exercise 6C

1. a) 20 cm^2 b) 14 cm^2 c) 70 mm^2 d) 7·5 m^2
2. a) 17·5 mm^2 b) 31·5 m^2
   c) 1140 mm^2 d) 252 m^2
3. 352 mm^2 4. 63 ft^2
5. a) 62100 cm^2 b) 14·4 m^2
6. a) a = 5 cm   b) h = 7 mm
   c) h = 5 m   d) d = 14 cm
7. a) 20 cm^2 b) 36 − 16 = 20 cm^2

Exercise 6D

1. a) 36 cm^2 b) 75 cm^2
   c) 1500 cm^2 d) 162 cm^2
2. a) 33·6 m^2 b) 2520 mm^2
   c) 2200 mm^2 d) 56·8 m^2
3. 5320 mm^2 4. 69 cm^2
5. a) 120·4 cm^2 b) 120·4 cm^2
6. 1932 mm^2
7. a) 620 mm, 400 mm   b) 248000 mm^2
8. 1 × 72, 4 × 18, 9 × 8, 16 × 4·5, 25 × 2·88, 36 × 2
9. 15 km
10. a) 0·5 unit^2 b) 2 unit^2 c) 8 unit^2 d) $\frac{1}{8}$
11. Pupils' own answers

Exercise 6E

1. a) 14 m^2 b) 24 mm^2 c) 300 cm^2
   d) 40 cm^2 e) 975 mm^2 f) 108 cm^2
   g) 119 m^2 h) 81 cm^2
2. Skyer 2625 cm^2 Flyhigh 3600 cm^2, Flyhigh
3. 1400 mm^2 4. 5·6 m^2

5 a) Pupils' own answers
   b) A = 40 cm^2, e.g. 8 × 10, 16 × 5, 20 × 4
6 a) 45 − 25 = 20 cm^2 b) 20 cm^2
7 Pupils' own answers, e.g. 6 × 8, 12 × 4, 16 × 3

Exercise 6F

1 a) 40 m^2 b) 15 cm^2 c) 31·5 m^2
   d) 380 mm^2 e) 137·5 cm^2 f) 780 mm^2
   g) 24·5 cm^2 h) 26·55 m^2
2 a) 9000 cm^2 b) 18000 ÷ 2 = 9000 cm^2
3 33·6 m^2 4 yes, 2565 km^2
5 21920 cm^2 6 10 m

Exercise 6G

1 a) 42 cm^2 b) 45 cm^2 c) 172 cm^2
   d) 210 cm^2 e) 48 m^2 f) 133 cm^2
2 Site A 368 m^2 , Site B 330 m^2 , Site A
3 a) 216 cm^2 b) 1375 cm^2 c) 384 m^2
4 a) 4000 cm^2 b) 33 m^2 c) 103·5 m^2
5 a) 61·5 m^2 b) £1960

Exercise 6H

1 a) 120 mm   b) 41 cm   c) 5·1 m
   d) 175 cm   e) 33·55 m   f) 60 cm
2 a) Site A 92 m   Site B 86 m
   b) Site A £3520 Site B £3280
   c) £240
3 a) 60 m   b) 120 m^2
4 a) 37·5 m^2 b) 23·84 m
5 a) 76 cm   b) 368 cm^2
6 a) 50 cm   b) 630 cm^2
7 84 cm
8 a) 3 m × 16 m   b) 12 cm × 27 cm
9 Pupils' own answers, e.g.
   square 4 cm × 4 cm, rectangle 6 cm × 3 cm

Exercise 6I

1 a) 108 cm^3 b) 1920 cm^3 c) 366 cm^3
   d) 372 m^3 e) 396 cm^3 f) 1075 m^3
2 a) 8 cm^3 b) 125 cm^3 c) 729 m^3
   d) 8000 mm^3

3 a) 24000 mm^3 b) 4320000 cm^3
   c) 37·5 m^3
4 a) a = 3 cm   b) b = 2 cm   c) c = 7 m
5 a) 250 m^3 b) 2336 m^3
6 a) Pupils' own answers, e.g.
      9 cm × 3 cm × 8 cm
   b) 6 cm × 6 cm × 6 cm

Exercise 6J

1 a) 180 cm^3 b) 840 mm^3 c) 1428 cm^3
   d) 21·15 m^3 e) 2790 cm^3 f) 11520 cm^3
2 Shape 1
   a) 27 cm^2,   b) 810 cm^3;
   Shape 2
   a) 34 cm^2,   b) 408 cm^2;
   Shape 3
   a) 1125 mm^2,   b) 135000 mm^3
3 a) 3   b) 20   c) 30
4 a) 800 cm^2 b) 11200 cm^3 c) 11·2 litres
5 a) 2·4 cm^2 b) 12 cm^3
   c) 231·6 g   d) £9727·20

Check-up

1

Rectangle	$A = l \times b$	Cuboid	$V = lbh$
Square	$A = l^2$	Cube	$V = l^3$
Triangle	$A = \frac{1}{2}bh$	Prism	$V = Ah$
Rhombus or kite	$A = \frac{1}{2}d_1d_2$		
Parallelogram	$A = bh$		
Trapezium	$A = \frac{1}{2}(a + b)h$		

2 H 67·5 cm^2 E 90 cm^2 C 100 cm^2
   T 108·5 cm^2 A 144 cm^2 R 150 cm^2
   E 169 cm^2 HECTARE. A hectare is an area of land
   equal to a square with 100 m sides (10000 m^2)
3 5·8 m^2, 5·25 m^2, Support 1 is larger
4 1000 cm^2, 1110 cm^2, 900 cm^2 Design 3 only
5 a) 3900 cm^2 b) 2760 cm^2 c) 1140 cm^2
6 No total area = 75 m^2 < 100 m^2
7 a) 131·7 m^2 b) 52·8 m
8 a) 40 cm   b) 42000 cm^3
   c) 36000 cm^3 d) 210 litres

9 a) triangular prism   b) $279\,mm^2$
   c) $23715\,mm^3$
10 a) $390000\,mm^2$ b) $97500\,mm^2$
   c) $150\,mm$

7 Expressions and equations

Exercise 7A

1 a) $4x$ b) $3y$ c) $3m$ d) k
2 a) $4x$ b) $6x$ c) $7y$ d) $11z$
   e) $9m$ f) $7w$ g) $3v$ h) $15h$
   i) $50g$ j) $7x$ k) $6q$ l) $15e$
3 a) $4x$ b) $5y$ c) p d) $2m$
   e) $6f$ f) $6t$ g) $6y$ h) $7d$
   i) $13l$ j) $15w$ k) $9j$ l) 0
4 a) $-2x$ b) $-4k$ c) $-z$ d) $-3t$
   e) $-5m$ f) $-6j$ g) $-3t$ h) $-5p$
   i) $-13x$ j) $-2h$ k) $-y$ l) $-4x$

Exercise 7B

1 a) $8x + y$ b) $7y + 3$ c) $4c + 1$
   d) $5h + 2$
2 a) $6x + 10y$ b) $6x + 8y$ c) $11m + 2n$
   d) $3k + 9p$ e) $10z + 2d$ f) $5x + 7$
   g) $13g + 6f$ h) $4t + 4u$
3 a) $9x + 3y + 8$ b) $14m + 5n + 4$
   c) $3r + 10s - p$ d) $4h + 13k + p$
   e) $-5x + 3y$ f) $-2p + 8q$
   g) $3r - 4t$ h) $-d - f + 7$
   i) $-5x - 5y - 6$
4 a) $5n - 6m - 2n + m$, $5n - 5m - 2n + m$,
      $10m - 6n - m$
   b) 0

Exercise 7C

1 a) $5x^2$ b) $8y^2$ c) $6m^2$ d) $5p^2$
   e) $7x^2$ f) $9y^2$
2 a) $8xy$ b) $10mn$ c) $4pq$ d) $2ab$
   e) $12uv$ f) $6xy$

3 a) $6x^2 + 7y^2$ b) $8p^2 + 7q^2$
   c) $6u^4 + 4v^3$ d) $6mn + 4pq$
   e) $15ab + 5cd$ f) $5pqr + rst$
4 a) $4m^2 - 2m$ b) $2x^2 - 12x$ c) $k^2 - k$
   d) $7xy - yz$ e) $12ab - 10cd$ f) $3abc - 2efg$
5 Ag – Silver, Pb – Lead, Fe – Iron, Ca – Calcium,
   Sn – Tin

6 a)
   | $30x - 8$ |
   | $19x - 3$ | $11x - 5$ |
   | $12x + 1$ | $7x - 4$ | $4x - 1$ |

   b)
   | $14x + 12$ |
   | $9x + 13$ | $5x - 1$ |
   | $6x + 9$ | $3x + 4$ | $2x - 5$ |

   c)
   | $12x^2 + 3$ |
   | $4x^2 + 1$ | $8x^2 + 2$ |
   | $x^2 + 1$ | $3x^2$ | $5x^2 + 2$ |

   d)
   | $88x + 64$ |
$44x + 19$	$44x + 45$
$19x + 4$	$25x + 15$	$19x + 30$
$7x + 3$	$12x + 1$	$13x + 14$	$6x + 16$
$2x + 4$	$5x - 1$	$7x + 2$	$6x + 12$	4

7 a)
   | $14m + 2$ |
   | $6m$ | $8m + 2$ |
   | $4m - 1$ | $2m + 1$ | $6m + 1$ |

   b)
   | $20pq + 5$ |
   | $8pq + 4$ | $12pq + 1$ |
   | $6pq + 1$ | $2pq + 3$ | $10pq - 2$ |

   c)
   | $7y^2$ |
   | $3y^2 + 3$ | $4y^2 - 3$ |
   | $y^2 + 2$ | $2y^2 + 1$ | $2y^2 - 4$ |

   d)
   | $29x + 17y$ |
$15x + 10y$	$14x + 7y$
$13x + 5y$	$5y + 2x$	$12x + 2y$
$13x + 2y$	$3y$	$2y + 2x$	$10x$
$12x + y$	$y + x$	$2y - x$	$3x$	$7x$

8 Pupils' own answers

9 a) LHS: 4 RHS: $5x$ b) LHS: 3 RHS: $7x^2$
   c) LHS: $13ab$ RHS: 4   d) LHS: $4q^2$ RHS: p^2, 5
   e) LHS: $3pq$ RHS: 3, $2pqr$, $3qr$, $5q$

Exercise 7D

1 a) 5   b) −4   c) 13   d) −2
   e) 10   f) 6   g) 22   h) 0
2 a) 12   b) 15   c) 12   d) 42
   e) −6   f) −18   g) 18   h) 180
3 a) 9   b) 7   c) 26   d) −4
   e) −8   f) 23   g) 29   h) −5
   i) −7   j) 11   k) 24   l) −2
4 a) 9   b) 0   c) 16   d) 2
   e) 34   f) 12   g) 9   h) 30
   i) 16   j) 2   k) 8   l) 18
5 a) Paris and Berlin   b) Pupils' own answers
6 a) 4   b) 25   c) 8   d) 625
   e) 32   f) 12   g) 36   h) 50
   i) 100   j) 32   k) 512   l) 750
7 a) 2   b) 7   c) 12   d) 3
8 a) 53   b) 5   c) 250   d) 516
9 Neptune, Mercury

Exercise 7E

1 a) 5   b) −1   c) −4   d) −3
   e) 5   f) 2   g) 0   h) −3
   i) −6   j) 5   k) −2   l) 6
2 a) −9   b) −30   c) −33   d) −20
   e) −25   f) −44
3 a) Card C   b) Card B
4 a) 2   b) −10   c) 2   d) −3
5

Column A	Column B	Column C
$m = 2, n = −3, p = 6$	$\frac{1}{2}$ of $mp + 6n$	−12
$m = −4, n = 10, p = −2$	$−8 + 3pm + 2m$	8
$m = 2, n = −1, p = 4$	$7mnp$	−56

6 a) 4   b) 12   c) 36   d) 9
   e) 45   f) 225   g) −12   h) −18
   i) 38   j) −8   k) −32   l) −512

7

¹1	²3	0		³1	3	⁴9
	1			4		0
	⁵1	0	⁶6	4		0
			9		⁷1	
⁸1	2	⁹5	2		4	
5		3		¹⁰6	4	¹¹8
0		¹²3	0	0		5

8 a) 2 and −2   b) 3 and −3
   c) 5 and −5   d) 10 and −10

Exercise 7F

1 a) $x = 4$ b) $x = 7$ c) $x = 3$ d) $x = 8$
   e) $x = 5$ f) $x = 16$ g) $x = 2$ h) $x = 19$
2 a) $x = 6$ b) $x = 12$ c) $x = 11$ d) $x = 14$
   e) $x = 13$ f) $x = 15$ g) $x = 13$ h) $x = 8$
3 a) $x = 6$ b) $x = 8$ c) $x = 1$ d) $x = 15$
   e) $x = 16$ f) $x = 9$ g) $x = 19$ h) $x = 13$
   i) $x = 12$ j) $x = 15$ k) $x = 9$ l) $x = 24$
   m) $x = 36$ n) $x = 30$ o) $x = 40$ p) $x = 0$
4 a) $x = −4$ b) $x = −7$ c) $x = −8$ d) $x = −3$
   e) $x = −11$ f) $x = −16$ g) $x = −24$ h) $x = −31$
   i) $x = 3$ j) $x = 8$ k) $x = −9$ l) $x = −16$
   m) $x = −5$ n) $x = −6$ o) $x = −1$ p) $x = −40$
5 a) $x = 4$ b) $x = 7$ c) $x = 3$ d) $x = 9$
   e) $x = 12$ f) $x = 15$ g) $x = 14$ h) $x = 67$
6 a) $x = −3$ b) $x = −11$ c) $x = −8$ d) $x = −4$
   e) $x = −9$ f) $x = −12$ g) $x = −50$ h) $x = −29$
7 a) $x = 5$ b) $x = 8$ c) $x = 7$ d) $x = 4$
   e) $x = 11$ f) $x = 9$ g) $x = 13$ h) $x = 18$

Exercise 7G

1 a) $x = 2$ b) $x = 4$ c) $x = 3$ d) $x = 5$
   e) $x = 7$ f) $x = 1$ g) $x = 3$ h) $x = 3$
   i) $x = 11$ j) $x = 25$ k) $x = 13$ l) $x = 20$

2 a) $x = 9$ b) $x = 7$ c) $x = 2$ d) $x = 7$
   e) $x = 9$ f) $x = 5$ g) $x = 30$ h) $x = 19$
   i) $x = 11$ j) $x = 10$ k) $x = 21$ l) $x = 12$
3 a) $x = -3$ b) $x = -2$ c) $x = -4$ d) $x = -7$
   e) $x = -6$ f) $x = -7$ g) $x = -1$ h) $x = -2$
   i) $x = -13$ j) $x = -3$ k) $x = -6$ l) $x = 11$
4 a) $x = 6$ b) $y = 5$ c) $p = 8$ d) $k = 2$
   e) $y = 9$ f) $m = 5$ g) $x = 2$ h) $p = 8$
   i) $m = 24$ j) $y = 8$ k) $r = 3$ l) $w = 34$
5 a) $x = 3$ b) $w = 2$ c) $k = 5$ d) $m = 6$
   e) $p = 7$ f) $n = 9$ g) $p = 3$ h) $x = 6$
   i) $q = 9$ j) $p = 12$ k) $x = 11$ l) $y = 60$
6 a) $x = -2$ b) $m = -5$ c) $p = -3$ d) $k = -13$
   e) $r = -40$ f) $q = -7$ g) $y = -12$ h) $g = -13$
   i) $x = -21$ j) $w = -42$ k) $k = -32$ l) $m = -9$

Exercise 7H

1 a) $x = 2$ b) $y = 5$ c) $m = 1$ d) $w = 4$
   e) $k = 6$ f) $q = 7$ g) $p = 15$ h) $h = 19$
   i) $k = 24$
2 a) $x = 3$ b) $m = 4$ c) $y = 5$ d) $k = 6$
   e) $x = 9$ f) $w = 12$ g) $p = 15$ h) $m = 8$
   i) $y = 21$
3 a) $r = 3$ b) $p = 2$ c) $x = 4$ d) $f = 9$
   e) $m = 7$ f) $y = 6$ g) $x = 5$ h) $q = 14$
   i) $m = 0$
4 a) $x = -5$ b) $m = -9$ c) $p = -8$ d) $q = -11$
   e) $w = -2$ f) $y = -4$ g) $n = -5$ h) $k = -7$
   i) $m = -6$
5 a) $x = 1$ b) $y = -1$ c) $w = -2$ d) $k = -4$
   e) $d = -3$ f) $m = 1$ g) $k = 4$ h) $t = -7$
   i) $p = -3$
6 a) $x = \dfrac{3}{2}$ b) $p = \dfrac{1}{4}$ c) $k = \dfrac{1}{2}$ d) $m = \dfrac{5}{2}$
   e) $p = \dfrac{3}{2}$ f) $y = -\dfrac{3}{5}$ g) $k = -\dfrac{1}{3}$ h) $x = -\dfrac{15}{2}$
   i) $y = -\dfrac{7}{3}$

Exercise 7I

1 a) i) $4x + 8 = 8x$
      ii) $x = 2$, dimensions rectangle: 2 cm,
           6 cm square: 4 cm
   b) i) $10x + 2 = 8x + 10$
      ii) $x = 4$, dimensions rectangle: 8 cm,
           13 cm parallelogram: 6 cm, 15 cm
   c) i) $10x + 15 = 15x$
      ii) $x = 3$, dimensions pentagon: 9 cm,
           triangle: 15 cm
   d) i) $8x + 6 = 9x$
      ii) $x = 6$, dimensions composite shape:
           9 cm, 5 cm, 13 cm, triangle: 16 cm,
           25 cm, 13 cm
2 Abdah 14; brother 20; sister 6
3 Grape 3 g; raspberry 1 g; strawberry 15 g
4 First: 90 m^2; Second: 144 m^2; Third: 126 m^2

Exercise 7J

1 a) 70   b) 60   c) 12
   d) 3   e) 5   f) 21
2 16 °C
3 a) £1200
   b) i) £1245   ii) £1338
   c) $W = 1200 + 0·18s$

Exercise 7K

1 a) 37   b) 79   c) 8   d) 19
2 a) 7   b) 6   c) 3   d) 36
3 a) £275   b) 20 packs   c) 40 packs
4 a) 4 cm   b) 2 cm

Check-up

1 a) $3x$ b) $4y$ c) $8m$ d) $3q$
   e) $10k$ f) $9p$ g) $-3m$ h) $-2r$
2 a) $8x + 3$ b) $3y + 2k$ c) $12m + 7n$
   d) $2p + 24q$ e) $8r + 5t$ f) $2m + 4n$
   g) $5u - 6v$ h) $10g - 2h$ i) $2a - 5b$
3 a) $11m + 5n + 2$ b) $2y + 4j + 2x$
   c) $12 + r + 16s$ d) $18u - 13v + 6w$
   e) $10p - 5q - 13$ f) $13x - 27y + 16k$

4 a) $8x^2$ b) $9y^2 + 2$
   c) $m^2 + 7m$ d) $2g^2 - 7h^2 + 10j^2$
   e) $14ab$ f) $2mn + 2n$
   g) $11pq - 11rt$ h) $6 + 5uv - 5xw$

5 a)
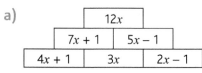
	12x	
7x + 1		5x − 1
4x + 1	3x	2x − 1

   b)
	8m + 6	
3m + 2		5m + 4
m + 2	2m	3m + 4

   c)
	$10x^2 + 2$	
$4x^2 + 3$		$6x^2 − 1$
$6 + x^2$	$3x^2 − 3$	$3x^2 + 2$

   d)
	22abc + 1	
12abc		10abc + 1
8abc − 3	3 + 4abc	6abc − 2

6 a) 12 b) 10 c) 25 d) 14 e) 1
   f) 18 g) 7 h) 58 i) 130 j) −70
   k) 36 l) 144

7 a) 4 b) −14 c) −4 d) 0 e) −24
   f) −12 g) −10 h) 17 i) 36 j) −18
   k) 24 l) −16

8 a) 4 b) 9 c) 12 d) 36 e) 4
   f) 2 g) 1 h) 26

9 a) $x = 1$ b) $y = 6$ c) $k = 5$ d) $w = 11$
   e) $q = 4$ f) $p = 12$ g) $y = 0$ h) $m = -7$
   i) $t = -1$ j) $v = -8$ k) $p = -10$ l) $m = -3$

10 a) $x = 6$ b) $k = 9$ c) $c = 3$ d) $h = 2$
   e) $p = 12$ f) $k = 3$ g) $x = 20$ h) $w = 8$
   i) $x = 2$ j) $y = 1$ k) $q = 2$ l) $m = 4$

11 a) $x = 3$ b) $y = 2$ c) $m = 8$ d) $p = 10$
   e) $h = 9$ f) $m = 14$ g) $k = 7$ h) $w = 9$

12 a) $x = -4$ b) $y = -14$ c) $m = -8$ d) $q = -16$
   e) $w = -7$ f) $k = -9$ g) $j = -11$ h) $r = -12$

13 a) $8x + 6 = 11x$
   b) $x = 2$, rectangle: 3 cm, 8 cm, triangle: 7 cm, 5 cm, 10 cm

14 a) 200 b) 39 c) 126
15 a) 16 b) 15
16 a) 8000 kg b) 9400 kg c) 300 bags

8 Money

Exercise 8A

1 a) 4 pack b) 10 eggs c) small loaf
   d) 2·5 kg e) 750 ml f) 1 kg
   g) 70 washes h) 7 apples i) 10 rolls
   j) Same price

2 a) No b) Yes c) No
   d) Same price e) Yes

3 200 g 4 500 ml 5 32p

6 a) £28·80 b) £57·60

Exercise 8B

1 a) Total = Outgoing = £192·45, Balance = £7·55
   b) Yes

2 a) Total = Outgoing = £530, Balance = £440
   b) £110 c) No he should save

3 a) Total = Outgoing = £474, Balance = £395
   b) £79

4 a) Total = Outgoing = £1145·50 Balance = £154·50
   b) Pupils' own c) $\frac{1}{5}$ d) 5%

5 Outgoing = £1054·50, Gig Tickets £45

6 a) £645 b) £480
   c) Food, Pupils' own answers

7 a) = SUM(B1:B4) b) = 4740/12
   c) = SUM(D3:D4) d) = F3 − F4
   e) Total = £537, Loan = £395 Incoming = £715, Balance = £178

8 a) = SUM(B1:B6) b) = SUM(D5:D6)
   c) = F5 − F6 d) £65·80
   e) £46·20

Exercise 8C

1

£970
£916
£796
£821
£584

2

£157·10
£135·10
£85·10
£1050·19
£1009·19
967·19

3

£1450
£1385·75
£1257·55
£1224·84
£646·84
£505·84

4

£605
£529·01
£517·01
£364·50
−£110·50
−£263·50
−£300·50

5 −£163 DR, −£596 DR, £2241, £123·54, £1441·17, £1380·12, £2821·29

Exercise 8D

1 a) £12   b) train

2 a) £1·90 b) £10   c) £4·50   d) 55p

3 a) £165   b) £270

4 a) £107·82   b) £47·94   c) £3·01

d) cancel anytime unlimited to try it

5 a) £390   b) £420   c) £395·88

d) Plus Link is least expensive

6 a) £2·60   b) Skate & sauna

c) £38   d) £360

e) Annual fee £300

7 a) Pupils' own answers, B is a good balance

b) £414   c) £270

8 a) Pupils' own answers, Heat Pump much better longer term

b) 3 years

Exercise 8E

1 £510   **2** £187·20   **3** £63

4 £803·40   **5** £243·60   **6** Maureen

7 a) Bonny Life   b) £112

8 Iona, Alba Bank;   Maisie, Society Bank; Richard, Scot Savers

9 £380   **10** 18 weeks   **11** 6 months

12 a) £195

b) £5 bonus per card, only spend in this shop, might lose the card

13 2·5%

Exercise 8F

1 a) £160   b) £800   c) £61

d) £324   e) £2400   f) £1440

2 Smile Finance, Happy Loanz, 2dayPay

3 a) Fife Credit Union

b) £780·60, £504·84, £550·80

c) £275·76

Exercise 8G

1 a) 875¥   b) €114   c) 8819₹

d) $127   e) 8159₽   f) 13969¥

2 a) 264570₹ b) 419070¥ c) 244770₽

d) $3810   e) 26250¥   f) €3420

3 a) 7875¥   b) €684   c) 61733₹

d) 279380¥   e) 14686·20₽ f) $63·50

g) 35000¥   h) €68·39   i) 22189₹

4 a) €1014·60   b) $793·75   c) ₽354916·50

d) 99180¥   e) 170187·50¥ f) 52473₹

5 1400000¥   **6** $984250   **7** 2381130₹

8 a) £571·43   b) £10·91   c) £314·96

d) £4385·96 e) £22·68   f) £6·44

g) £367·69   h) £787401·57 i) £11339·15

j) £322·81   k) £523·54   l) £629·91

9 £157894·74 10 £349344·98 11 £9257·14

12 a) £9000   b) £45 per gram

13 £214·29   14 £51·75

15 a) €1026

    b) total = outgoing = €820 balance = €206

    c) £180·70

16 a) China £615·77, USA £551·18, £539 UK,
       France £557·89

    b) Chinese yuan   c) pounds sterling

17 a) £129·41   b) £1950

18 a) Bank has best rate b) €425

Check-up

1 a) 4kg bag   b) 300 ml

2 80 washes

3 Pupils' own answers, balance ≥ 0

4 a) yes £2 less per hour b) £4

   c) £1·33   d) 1 hour

5 a) £51   b) £1684

6 £836·40

7 a) Loans4cars   b) £4200

   c) £14400   d) 10 years   e) no

8 £1930·50, £1405·50, £1365·50, £1290·50,
   £1450·50, £1398·50

9 a) HK$99·20 b) HK$992   c) HK$4960

   d) 1509000₩ e) 1358100₩ f) 66300 kr

   g) £200   h) £10

10 Norwegian krone

11 a) 2435 zł

    b) total = outgoing = 1614 zł, balance = 821 zł

    c) £168·58

9 Patterns and relationships

Exercise 9A

1 a) 14, 17, 20 'start at 2 and add 3 each time'

   b) 33, 39, 45 'start at 9 and add 6 each time'

   c) 53, 66, 79 'start at 1 and add 13 each
      time'

   d) 17, 15, 13 'start at 25 and subtract 2 each
      time'

   e) 31, 23, 15 'start at 63 and subtract 8 each
      time'

   f) 44, 30, 16 'start at 86 and subtract 14 each
      time'

   g) −10, −16, −22 'start at 14 and subtract
      6 each time'

   h) 17, 29, 41 'start at −19 and add 12 each
      time'

   i) 56, 37, 18 'start at 113 and subtract
      19 each time'

   j) −27, −35, −43 'start at −3 and subtract
      8 each time'

   k) 55, 69, 83 'start at 13 and add 14 each
      time'

   l) 10·9, 12·2, 13·5 'start at 7 and add 1·3
      each time'

2 a) 32, 64, 128 'start at 2 and multiply by
      2 each time'

   b) 81, 243, 729 'start at 1 and multiply by
      3 each time'

   c) 250, 1250, 6250 'start at 2 and multiply by
      5 each time'

   d) 50, 25, 12·5 'start at 400 and divide by
      2 each time'

   e) 45, 15, 5 'start at 3645 and divide by
      3 each time'

   f) 70, 7, 0·7 'start at 70000 and divide by
      10 each time'

   g) 500, 2500, 12500 'start at 4 and multiply
      by 5 each time'

   h) 36, 18, 9 'start at 288 and divide by 2 each
      time'

   i) 189, 567, 1701 'start at 7 and multiply by
      3 each time'

   j) 25, 5, 1 'start at 3125 and divide by 5 each
      time'

   k) 80, 20, 5 'start at 5120 and divide by
      4 each time'

   l) 256, 1024, 4096 'start at 4 and multiply by
      4 each time'

3 73   4 14   5 15   6 154

7 8, 13, 21, 34, 55

Exercise 9B

1 1, 4, 9, 16, 25, 36, 49, 64, 81, 100

2 The differences are the sequence of increasing positive odd numbers

3 121, 144, 169, 196, 225, 256, 289, 324, 361, 400

4 Sharon is correct. Multiplying two odd numbers always results in an odd number (as an odd number does not have a factor of two).

5 $6^2 + 8^2 = 10^2$, $9^2 + 12^2 = 15^2$, $12^2 + 16^2 = 20^2$, $5^2 + 12^2 = 13^2$, $8^2 + 15^2 = 17^2$

Exercise 9C

1

2 1, 3, 6, 10, 15, 21, 28, 36, 45, 55

3 a) 4   b) 8   c) 10   d) 54

4 The sum of the nth and $(n + 1)$th triangular numbers is the $(n + 1)$th square number.

   For example, the 3rd and 4th triangular numbers give a 4 by 4 square:

5 a) 144   b) 400

6 No. Adding the 3rd square and triangular numbers gives 15 (not prime).

Exercise 9D

1 a) yes, 6   b) no   c) yes, −10 d) no
   e) yes, 8   f) no   g) yes, $\frac{2}{7}$ h) yes, $\frac{2}{5}$
   i) no   j) yes, 12

2 a) 5   b) −2   c) 8   d) 20
   e) 3   f) −4   g) −3   h) 1
   i) 8·5   j) 11·8 k) $-\frac{13}{5}$ l) $\frac{5}{4}$

3 a) 21, 81   b) 22, 112   c) 20, 65
   d) 56, 191   e) 31, 136   f) 16, 91
   g) −39, −174 h) −55, −265 i) 10·5, 18
   j) 11, 8   k) −1, 5   l) $\frac{17}{4}, \frac{31}{2}$

4 a) i) 9, 13, 17, 21   ii) 5, 11, 17, 23
      iii) −1, −6, −11, −16

   b) The common difference is the number multiplying n.

5 a) 28, 35, 42, 7n   b) 12, 15, 18, 3n
   c) 48, 60, 72, 12n   d) −16, −20, −24, −4n
   e) 11, 13, 15, 2n + 3   f) 18, 22, 26, 4n + 2
   g) 16, 19, 22, 3n + 4   h) 37, 44, 51, 7n + 9
   i) 11, 14, 17, 3n − 1   j) 35, 46, 57, 11n − 9
   k) 30, 39, 48, 9n − 6   l) 16, 21, 26, 5n − 4
   m) −4, −7, −10, −3n + 8 n) −1, −5, −9, −4n + 15
   o) −13, −16, −19, −3n − 1
   p) −28, −33, −38, −5n − 8

Exercise 9E

1 a)

   b) 13, 16, 19, …, 31   c) m = 3s + 1
   d) 76   e) 34

2 a)

   b) 9, 11, 13, …, 21   c) m = 2t + 1
   d) 37   e) 23

3 a) 39·2, 41   b) F = 1·8C + 32
   c) 68 °F   d) 10 °C   e) 32 °F

4 a) 720°, 900°, 1080° b) a = 180s − 360
   c) 1800°   d) 20
   e) Answer is 0: can't have a 2-sided polygon

5 Pupils' own graphs

Check-up

1 a) 24, 29, 34 'start at 4 and add 5 each time'
   b) 16, 13, 10 'start at 28 and subtract 3 each time'
   c) 24, 48, 96 'start at 3 and multiply by 2 each time'
   d) −21, −33, −45 'start at 15 and subtract 12 each time'
   e) 16, 8, 4 'start at 128 and divide by 2 each time'

f) 21·8, 25·5, 29·2 'start at 7 and add 3·7 each time'

g) 8, $7\frac{1}{4}$, $6\frac{1}{2}$ 'start at 11 and subtract $\frac{3}{4}$ each time'

h) −128, 512, −2048 'start at 2 and multiply by −4 each time'

2 −30   3 19   4 45   5 20

6 16 and 64   7 3 and 21   8 138

9 a) 16   b) 1   c) −14   d) 3

10 a) 24, 412   b) 19, 892

c) −34, −1004   d) −15, −888

11 a) 210   b) 300   c) 5050

12 a) 41, 48, 55, $7n + 13$

b) 26, 33, 40, $7n - 2$

c) 196, 246, 296, $50n - 4$

d) 9, 9·5, 10, $0·5n + 7$

13 a) 3, −1, −5, $-4n + 19$

b) 0, −2, −4, $-2n + 8$

c) −11, −13, −15, $-2n - 3$

d) −49, −60, −71, $-11n - 5$

14 a)

b) 17, 21, 25, …, 41   c) $4n + 1$

d) 93   e) 19th pattern

15 a)
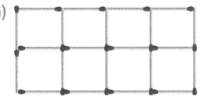

b) 22, 27, 32, …, 52   c) $m = 5n + 2$

d) 97   e) 25th pattern

16 a) £8   b) £19

c) $c = 8d + 19$ d) 10 GB

17 a) boards: 49, 60 screws: 96, 120

b) $w = 11u + 5, s = 24u$

c) Correct number of boards but 3 screws too few.

11 Coordinates, symmetry and scale

Exercise 11A

1 a) A(3, 5), B(7, 10), C(8, 3), D(1, 7), E(6, 0), F(10, 5), G(6, 7), H(0, 2)

b) E and G   c) A and F, D and G

2 Pupils' own diagrams

3 a) Diagram   b) Diagram   c) (4, 6)

4 a) (5, 4)   b) (5, 6)   c) (7, 3)   d) (5, 7)

5 They have the same x-coordinate.

6 They have the same y-coordinate.

7 a) Axes start at 1, not 0

b) Uneven spacing on axes

c) Numbers are not on lines

d) Numbers are on the wrong side of axes

8 a) Pupils' own diagrams   b) D(6, 5)

9 a) Pupils' own diagrams   b) (8, 5) or (0, 5)

10 a) Pupils' own diagrams   b) D(8, 4)

11 a) Pupils' own diagrams   b) D(0, 3)

Exercise 11B

1 a) A(−8, 2), B(−4, 0), C(0, 6), D(7, 2), E(−7, −7), F(3, −3), G(−2, 4), H(−9, −3), I(0, −7), J(9, −4), K(−5, 9), L(−2, −3), M(5, 8), N(1, 2), O(−7, 7)

b) H, L, F   c) D   d) A, N, D and H, A, O

2 Check pupils' diagrams

3 a) Trapezium   b) integers

c) quadrilateral   d) Hypatia

e) Fibonacci   f) Alan Turing

4 a) (−9, 6), (−3, 5), (−5, 8), (5, −9), (−9, 6)

b) (−3, −2), (−3, 5), (−5, 8), (4, 0), (8, −6), (3, −3), (−5, 8), (5, −9)

c) (3, −3), (−9, 6), (8, −6), (−3, −2), (5, −9), (1, −9), (−3, 5), (−3, −2), (6, −4), (5, −9), (−9, 6)

d) (0, −6), (2, 2), (−3, −2), (−3, −2), (5, −9), (4, 0), (−9, 6), (2, 2)

5 Pupils' own responses

Exercise 11C

1. a) rectangle   b) kite
   c) rhombus   d) square
   e) trapezium   f) isosceles triangle
   g) pentagon   h) parallelogram
2. a) 4, 4, 4   b) 2, 2, 4   c) 2, 4, 2
   d) 1, 2, 1   e) 0, 2, 2
3. a) 2   b) 3   c) 1   d) 0
   e) 1   f) 2   g) 8   h) 1
4. a) 3, 5, 6, 8   b) 7   c) 10
5. Pupils' own diagrams

Exercise 11D

1. Pupils' own diagrams   2. Pupils' own diagrams
3. Pupils' own diagrams   4. Pupils' own diagrams

Exercise 11E

1. Pupils' own diagrams
2. a) 4   b) 32 cm
3. a) 3   b) 27 cm
4. a) 4   b) 72 cm   c) 3744 cm^2

Exercise 11F

1. Pupils' own diagrams
2. a) $\frac{1}{4}$ b) 7 cm
3. a) $\frac{1}{7}$ b) 15 cm
4. a) $\frac{2}{3}$ b) 16 cm
5. a) height = 15 cm, base = 24 cm   b) 180 cm^2
6. a) 4 cm   b) 25 times

Exercise 11G

1. Pupils' own diagrams
2. a) Pupils' own diagrams
   b) 0·446 seconds
3. a) Pupils' own diagrams   b) 883
   c) e.g. space needed for cars to drive in and out

4. a) 9 cm by 5·4 cm   b) diagram
5. a) Pupils' own diagrams
   b) 15 cm   c) 150 m
6. Pupils' own diagrams
7. a) 1 cm : 2 m   b) 30 m
8. a) 1 cm : $\frac{1}{3}$ m   b) 9 m
9. a) 1 cm : 0·5 m   b) 12·3 m

Check-up

1. A(6, 2), B(2, −7), C(−3, 8), D(−5, −10), E(9, −2),
   F(−7, 0), G(0, −4), H(5, 7)
2. a) Pupils' own diagrams   b) D(3, 1)
3. Pupils' own diagrams
4. a) Pupils' own diagrams   b) D(6, 2)
5. a) Pupils' own diagrams   b) D(10, 6)
   c) (6, 6)
6. Pupils' own diagrams
7. a) 2   b) 0   c) 1   d) 3
8. Pupils' own diagrams
9. Pupils' own diagrams
10. a) Pupils' own diagrams
    b) 14·3 cm   c) 85·8 feet
11. Pupils' own diagrams
12. a) 1 cm : 20 feet   b) 1060 feet

12 Data analysis

Exercise 12 A

1. Categorical: type of pet, city, sport, career
   Numerical: time, calories, share price, length,
   age, test score, wages, savings, blood pressure
2. a) survey, categorical
   b) research data, numerical
   c) survey, categorical
   d) gather data, numerical

Exercise 12B

1
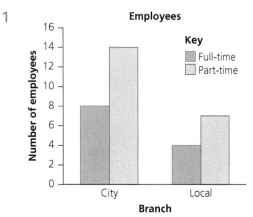

2
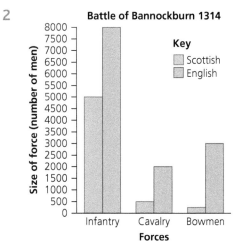

3 a)
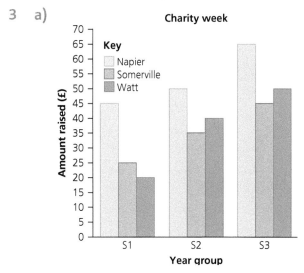

    b) S3   c) Napier

4 a) Hatchback   b) Petrol   c) 5

    d) $\dfrac{1}{5}$ e) $\dfrac{1}{4}$

5 a) 110   b) Sat matinee   c) $\dfrac{1}{3}$

    d) Sat evening   e) upper circle

6 a) £1000   b) 80%

    c) Overall sales are down £1000, proportion of food and drink vouchers much smaller

    d) $\dfrac{1}{8}$

7 a)
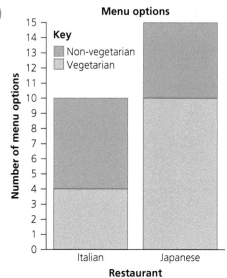

    b) Japanese, more options and greater proportion of options.

    c) Italian 40% Japanese $66\dfrac{2}{3}$%

8 a) $\dfrac{1}{3}$

    b) She didn't visit any countries in Africa.

    c) Lizzie travelled less in Europe but more in Asia and Africa, and more overall.

Exercise 12C

1 a)
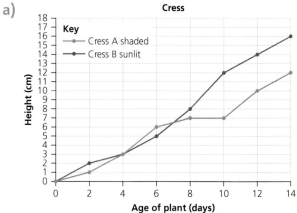

    b) 10 cm   c) cress A

    d) day 4, day 7   e) didn't grow

    f) sunlight helps growth

2 a)

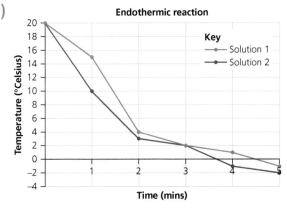

Endothermic reaction

Key
—•— Solution 1
—•— Solution 2

Temperature (°Celsius) / Time (mins)

b) 0, 3 minutes

c) Solution 2, 3 minutes 45 seconds

d) draws in heat, cooling

e) Solution 1 dropped 11°C, 1–2 minutes

3 a) £26   b) 14:04   c) 14:00

d) rising and falling often

e) steady rise, drop at 14:15   f) £5000

g) £7500

4 a) 56%   b) there was a small drop, 1%

c) 1990   d) 24%   e) 2000 – 2005

f) Hepatitis (84%) vaccinations almost
caught up with measles (85%)

Exercise 12D

1 a) 100   b) 50   c) 40   d) 10

2 a) 100   b) 200   c) 120   d) 180

3 a) 90   b) 225   c) 360   d) 18°   e) 180

4 a) 240   b) 30   c) 130

5 a) £14·4 billion   b) £17·6 billion

c) £48 billion

d) whisky, salmon, beef and lamb, chemicals,
petroleum products, electronics

Exercise 12E

1 a)

	Hours	Relative frequency	Angle at centre
Cardio	12	$\frac{12}{24} = \frac{1}{2}$	$\frac{1}{2}$ of 360° = 180°
Strength	8	$\frac{8}{24} = \frac{1}{3}$	$\frac{1}{3}$ of 360° = 120°
Dynamic	4	$\frac{4}{24} = \frac{1}{6}$	$\frac{1}{6}$ of 360° = 60°
Total	24		

b)

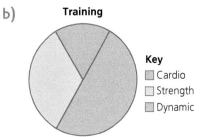

Training

Key
▢ Cardio
▢ Strength
▢ Dynamic

2 a)

	Amount raised (£)	Relative frequency	Pie chart angle
Bag packing	400	$\frac{400}{1200} = \frac{1}{3}$	$\frac{1}{3}$ of 360° = 120°
Talent show	360	$\frac{360}{1200} = \frac{3}{10}$	$\frac{3}{10}$ of 360° = 108°
Fun run	240	$\frac{240}{1200} = \frac{1}{5}$	$\frac{1}{5}$ of 360° = 72°
Smoothie bike	200	$\frac{200}{1200} = 6$	$\frac{1}{6}$ of 360° = 60°
Total	1200		

b) Charity fundraising

Key
▢ Bag packing
▢ Talent show
▢ Fun run
▢ Smoothie bike

3 a)

	Frequency	Relative Frequency	Angle at centre
Param	15	$\frac{15}{33}$	$\frac{15}{33} \times 360 = 164°$
Jodie	7	$\frac{7}{33}$	$\frac{7}{33} \times 360 = 76°$
Hector	11	$\frac{11}{33}$	$\frac{11}{33} \times 360 = 120°$
Total	33		

b) Pupil Council

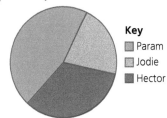

Key
▢ Param
▢ Jodie
▢ Hector

4 **Scotland's electricity**

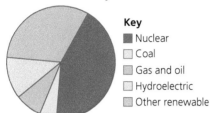

Key
- ■ Nuclear
- □ Coal
- ▨ Gas and oil
- ▨ Hydroelectric
- ▨ Other renewable

Nuclear	43%	155°
Coal	5%	18°
Gas and oil	8%	29°
Hydroelectric	12%	43°
Other renewable	32%	115°

Exercise 12F

1 a) wages b) time spent with family

   c) height

2 a)

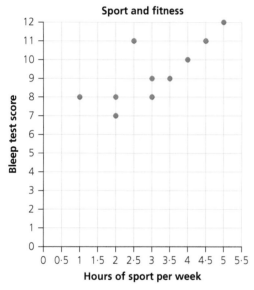

   b) There is a positive correlation. People who do more sports score higher on the beep test.

3 a)

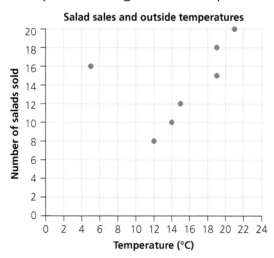

   b) There is positive correlation. More salads are sold on hotter days.

   There is an outlier. On a 5 °C day, 16 salads were sold.

4 a)

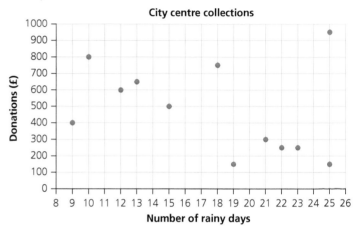

   b) There is a negative correlation. As the number of rainy days increases the total donated decreases. December is an outlier, it has 25 rainy days but the highest total donations.

   c) We don't know how the link works. Maybe there are fewer people at the shops or the volunteers are standing in a less visible spot. Maybe it is just a coincidence. Scatter graphs do not prove one thing causes another. They only provide evidence of where connections might exist.

Exercise 12G

1 Unreliable, few people would take time to email. Actually, 10% taking time to say they didn't like it seems high.

2 a) No, too small.

   b) No, pets at the puppy parlour are likely to be well cared for.

3 a) For example, pupils at front of queue likely to get their first choice and be happy not to wait.

   b) For example, pupils who go home won't be included. Pupils might be from same year group.

   c) Assign all pupils a number and use a random number generator.

4 Leading question. Pupils had no chance to say what made them unhappy. Sample size too small.

5 The scale is very large. The real difference in the scores is only 0·5% and 1%. The axis is drawn at 84% making it look like the girls' spelling scores fall below something. The girls' scores are actually very high. The choice of pink is stereotypical.

6 The graph shows no correlation. Scatter graphs can never prove a causal link.

7 The sample is too small. Pupils who did well may be biased. They might not be honest with the head of department. The slices have different areas and there is no scale.

8 The scale on the y-axis is going up in powers of 4. This kind of scale can be useful to display a wide range of numbers on a compact scale but here it might disguise a huge rise in customer complaints

9 Pupils' own diagrams

Exercise 12H

1 a) Bar chart or pie chart

b) Data is categorical

2 a) Cara needs to edit/add the title and axes labels. There are no numbers on the axes. Cara should add a key.

b) Data is numerical

3 a) William needs to edit/add the key. His chart only shows Port Glasgow's data.

b)
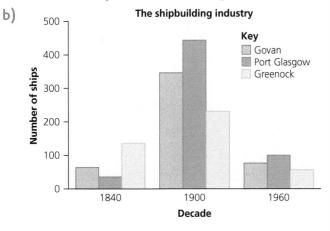

4 a)
Memory test

b) There is a negative correlation. Older people remember fewer items. There is an outlier though, a 61 year old remembered 18 items.

5

Type of graph or chart
Pie Chart
Compound Bar Graph
Compound Line Graph
Scatter Graph

6 Pupils' own answers

Check-up

1

Categorical	Numerical
Eye colour	Price
Hobby	Flight time
Exam subject	Exam length

2 a) 10   b) 30   c) 50%

d) Yes, his total time dropped, but he spent a little more time on his browser.

3 a) 6000   b) 3000   c) $\frac{2}{3}$

d) Fewer steps overall and a smaller proportion in the morning.

4 a) 20

b) The numbers were steady at around 20 most of the night. At 8 pm numbers fell to just 1 person and at the end of the night numbers went up to 28.

c) 14   d) Steadily increasing all night

e) 6 people

f) The ceilidh band got more people dancing early in the night but from 8:45 pm the DJ did better. Overall, the DJ got more people dancing.

5 a) $\frac{1}{4}$ b) 15   c) $\frac{2}{3}$ d) 40   e) 5

6 a)

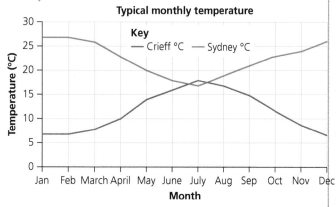

Typical monthly temperature

b) 11 °C   c) 10 °C

d) Much higher overall. High Dec/Jan and low June/July. Crieff is the opposite.

e) July

7 Sample size is too small and can't assume a causal link.

13 Angles

Exercise 13A

1 a) ∠ABC or ∠CBA   b) ∠DFE or ∠EFD
   c) ∠HGI or ∠IGH   d) ∠LKJ or ∠JKL

2 a) ∠DAB, right   b) ∠ABC, right
   c) ∠GDE, obtuse   d) ∠EFG, obtuse
   e) ∠KHI, obtuse   f) ∠IJK, acute
   g) ∠JKH, acute   h) ∠NLM, acute
   i) ∠LMN, obtuse   j) ∠LNM, acute
   k) ∠SOP, obtuse   l) ∠PQR, obtuse
   m) ∠QRS, obtuse   n) ∠WTU, acute
   o) ∠TUV, obtuse   p) ∠UVW, acute
   q) ∠VWT, obtuse

Exercise 13B

1 a) ∠ABC = ∠EBD   b) ∠GJH = ∠FJI
      ∠ABE = ∠CBD   ∠GJF = ∠HJI

c) ∠RLN = ∠KLM
   ∠ONP = ∠LNQ

2 a) 35°, 145°, 35°   b) 122°, 58°, 122°
   c) 17°, 163°, 17°   d) 90°, 28°, 62°, 28°
   e) 120°, 60°, 120°, 78°, 102°, 78°
   f) 28°, 112°, 28°

Exercise 13C

1 a) ∠BAE and ∠CED   b) ∠FJG and ∠JIH
   c) ∠KLO and ∠LMN

2 Pupils' own diagrams

3 Pupils' own diagrams

4 Since ∠BAC ≠ ∠DCE

5 a) 77°, 103°, 134°, 46°
   b) Angles are equal (triangles are similar)

6 5a = 180
   a = 36°

7 No, since the angles under the shelves are 60° and 59° (i.e. the shelves are not parallel)

Exercise 13D

1 a) ∠ABC and ∠BCD   b) ∠QPR and ∠PRS
   c) ∠VSU and ∠SUT

2 Pupils' own diagrams

3 Pupils' own diagrams

4 a) 5x = 180°   b) 4x = 180°   c) 36x = 180°
      x = 36°   x = 45°   x = 5°
      4x = 144°   3x = 135°   23x = 115°
                                    13x = 65°

5 AB and EF since alternate angles are equal.

Exercise 13E

1 a) x = 110°   b) x = 63°   c) x = 49°
   d) x = 37°   e) x = 47°   f) x = 38°
   g) x = 29°   h) x = 100°   i) x = 60°

2 a) x = 55°, y = 125°
   b) x = 114°, y = 85°
   c) x = 39°, y = 141°, z = 129°
   d) x = 103°, y = 45°, z = 135°
   e) a = 164°, b = 136°, c = 120°
   f) x = 39°, y = 39°, z = 141°
   g) x = 94°, y = 43 , z = 43°

h) $a = 86°, b = 94°, c = 94°, d = 36°$

i) $x = 25°, y = 112°, z = 16°$

3 a) $9x = 180°$ b) $12x = 180°$ c) $18x = 180°$
      $x = 20°$ $x = 15°$ $x = 10°$
      $2x = 40°$ $2x = 30°$ $4x = 40°$
      $3x = 60°$ $5x = 75°$ $7x = 70°$
      $4x = 80°$

   d) $3x = 180°$ e) $30x = 180°$ f) $60x = 180°$
      $x = 60°$ $x = 6°$ $x = 3°$
                       $4x = 24°$ $4x = 12°$
                       $10x = 60°$ $5x = 15°$
                       $16x = 96°$ $51x = 153°$

4 No: remaining angle must be 76°.

5 Both 45°

6 An obtuse angle is greater than 90°, so two obtuse angles are greater than 180°.

7 $\angle DCE = 80°$, $\angle CAD = 20°$, $\angle ABE = 60°$, $\angle BEA = 60°$, $\angle EAB = 60°$, $\angle BEC = 120°$, $\angle ECB = 20°$

Exercise 13F

1 a) $x = 98°, y = 38°$

   b) $a = 45°, b = 135°, c = 45°, d = 135°, e = 135°$

   c) $a = 63°, b = 117°, c = 63°$

   d) $a = 39°, b = 112°, c = 29°, d = 112°$

   e) $a = 57°, b = 55°, c = 55°, d = 123°, e = 68°$

   f) $a = 61°, b = 31°, c = 31°$

   g) $a = 124°, b = 62°, c = 62°, d = 118°$

   h) $p = 61°, q = 119°, r = 119°, s = 61°, t = 119°, u = 61°, v = 29°$

   i) $a = 85°, b = 44°, c = 41°, d = 139°$

   j) $a = 45°, b = 45°, c = 135°, d = 135°$

   k) $p = 71°, q = 79°, r = 11°$

   l) $a = 22°, b = 136°, c = 224°, d = 60°, e = 38°, f = 38°$

   m) $a = 124°, b = 56°, c = 33°, d = 33°, e = 147°, f = 147°, g = 147°, h = 33°$

   n) $a = 29°, b = 151°, c = 60°, d = 60°, e = 60°, f = 165°, g = 16°$

   o) $a = 107°, b = 37°, c = 143°, d = 73°, e = 37°, f = 143°, g = 107°$

2 a) $60x = 180°, x = 3°$

   b) $18x + 90 = 360°, x = 15°$

   c) $23x - 4 = 180°, x = 8°$

   d) $5x - 10 = 180°, x = 38°$

   e) $5x + 5 = 180°, x = 35°$

   f) $7x + 5 = 180°, x = 25°$

Exercise 13G

1 a) $x = 65°$ b) $x = 25°$
   c) $x = 137°$ d) $x = 216°$
   e) $x = 147°$ f) $x = 81°$

2 a) $8x + 120 = 360°, x = 30°$
   b) $18x + 90 = 360°, x = 15°$
   c) $5x + 160 = 360°, x = 40°$

Exercise 13H

1 a) 60°   b) 60°   c) 120°   d) 60°

2 a) 135°, 45°   b) 144°, 36°
   c) 162°, 18°   d) 150°, 30°
   e) 140°, 40°   f) 160°, 20°
   g) 176·4°, 3·6°   h) 128·57°, 51·43°

Exercise 13I

Pupils' own diagrams

Exercise 13J

Pupils' own diagrams

Exercise 13K

Pupils' own diagrams

Exercise 13L

Pupils' own diagrams

Exercise 13M

1 Pupils' own diagrams

2 a) 090°   b) 270°   c) 180°   d) 045°
   e) 225°   f) 000°   g) 135°   h) 315°

3 a) 073°   b) 030°   c) 115°   d) 145°
   e) 260°   f) 320°

4 a) 240°   b) 228°   c) 273°   d) 293°
   e) 085°   f) 112°

5 a) 220°   b) 215°   c) 284°   d) 076°
   e) 120°   f) 197°   g) 054°   h) 288°

6 Pupils' own diagrams

7 a) 230°   b) 250°   c) 300°   d) 355°
   e) 116°   f) 013°   g) 171°   h) 187°

8 Pupils' own diagrams

9 Pupils' own diagrams

10 a) Pupils' own diagrams

   b) 11·9 km   c) 239°

11 Approx. 770 km

12 a) Pupils' own diagram

   b) 4·8 km   c) 294·4°

13 a) Pupils' own diagram

   b) 26·5 m   c) 43·6 m   d) 20 m

Check-up

1 a) ∠PQR or ∠RQP   b) ∠YXZ or ∠ZXY

   c) ∠ABC or ∠CBA   d) ∠RSP or ∠PSR

2 a) 180°   b) 360°   c) 360°

   d) alternate   e) corresponding

3 a) $a = 102°, b = 78°$

   b) $c = 148°, d = 32°, e = 148°$

   c) $f = 51°, g = 129°, h = 309°$

   d) $i = 121°$ e) $j = 79°, k = 20°$ f) $l = 83°$

4 a) $9x = 180°$ b) $3x = 180°$

       $x = 20°$ $x = 60°$

      $2x = 40°$ $x + 10 = 70°$

      $7x = 140°$ $x - 10 = 50°$

   c) $5x + 15 = 180°$ d) $13x + 100 = 360°$

       $x = 33°$ $x = 20°$

      $2x = 66°$ $5x + 4 = 104°$

     $3x + 15 = 114°$ $8x + 6 = 166°$

   e) $5x + 5 = 180°$ f) $8x + 80 = 360°$

       $x = 35°$ $x = 35°$

     $x + 2 = 37°$ $3x + 1 = 106°$

    $3x + 3 = 108°$ $2x + 4 = 74°$

                            $3x + 13 = 118°$

5 a) $a = 65°, b = 48°, c = 115°, d = 132°, e = 67°$

   b) $a = 31°, b = 126°, c = 149°, d = 54°, e = 31°$

   c) $a = 68°, b = 44°, c = 68°, d = 68°, e = 112°,$
      $f = 112°$

   d) $a = 105°, b = 25°, c = 48°$

   e) $a = 78°, b = 24°, c = 102°, d = 58°$

   f) $a = 82°, b = 41°, c = 44°, d = 38°, e = 56°$

6 Pupils' own diagrams   7 Pupils' own diagrams

8 Pupils' own diagrams   9 Pupils' own diagrams

10 Pupils' own diagrams

11 a) 120°, 60°   b) 135°, 45°   c) 156°, 24°

12 a) East   b) South-east

   c) North-west   d) North-east

13 a) 075°   b) 105°   c) 310°

14 Pupils' own diagrams

15 Check pupils' diagrams

16 a) Check pupils' diagrams

   b) 71 km   c) 5·45 hours (5 hours 27 mins)

14 Probability

Exercise 14A

1 a) even chance   b) very unlikely

   c) likely   d) certain

   e) unlikely   f) unlikely

2 a) 4   b) very unlikely   c) 4

3 a) unlikely   b) even chance

   c) very unlikely   d) impossible

   e) even chance   f) even chance

Exercise 14B

1 a) $\frac{1}{3}$ b) $\frac{2}{3}$ c) 0

2 a) $\frac{3}{5}$ b) $\frac{2}{5}$ c) $\frac{3}{8}$

3 a) $\frac{2}{11}$ b) $\frac{7}{11}$ c) $\frac{4}{11}$

4 a) $\frac{1}{4}$ b) $\frac{1}{13}$ c) $\frac{3}{13}$ d) $\frac{5}{13}$

   e) $\frac{5}{13}$ f) $\frac{1}{26}$ g) $\frac{4}{13}$ h) $\frac{3}{13}$

5 a) $\frac{1}{6}$ b) $\frac{1}{3}$ c) $\frac{2}{3}$

   d) $\frac{1}{2}$ e) $\frac{1}{6}$

6 a) $\frac{1}{2}$ b) $\frac{3}{5}$ c) $\frac{9}{13}$ d) 0·3

   e) 0·85   f) 0

7 a) $\dfrac{1}{8}$ b) $\dfrac{1}{4}$ c) $\dfrac{3}{8}$ d) $\dfrac{3}{4}$

   e) $\dfrac{1}{2}$ f) $\dfrac{7}{8}$ g) $\dfrac{7}{12}$ h) $\dfrac{17}{24}$

8 a) $\dfrac{2}{35}$ b) $\dfrac{5}{7}$ c) $\dfrac{23}{35}$ d) $\dfrac{34}{69}$

   e) $\dfrac{35}{69}$ f) $\dfrac{20}{69}$

9 a) $\dfrac{5}{12}$ b) $\dfrac{1}{4}$ c) $\dfrac{7}{12}$ d) $\dfrac{1}{12}$

10 a) $\dfrac{5}{24}$ b) $\dfrac{1}{6}$ c) $\dfrac{11}{24}$ d) $\dfrac{7}{22}$

    e) $\dfrac{1}{2}$ f) $\dfrac{2}{11}$

11 a) $\dfrac{5}{12}$ b) $\dfrac{1}{2}$ c) $\dfrac{1}{3}$

12 a) $\dfrac{4}{15}$ b) $\dfrac{1}{15}$ c) 0

    d) $\dfrac{2}{5}$ e) $\dfrac{3}{10}$ f) $\dfrac{7}{10}$

13 a) $\dfrac{5}{6}$ b) lower

14 a) $\dfrac{6}{35}$ b) $\dfrac{9}{35}$ c) $\dfrac{19}{70}$

    d) $\dfrac{8}{35}$ e) $\dfrac{13}{14}$

15 a) 7 b) 6 c) $\dfrac{1}{6}$

    d) $\dfrac{1}{2}$ e) $\dfrac{1}{6}$

Exercise 14C

1 a) 5 b) 25 c) 41
2 a) 15 b) 50 c) 1
3 a) 5 b) 13 c) 75
4 a) 5 b) 105 c) 28
5 a) 9 b) 12 c) 21 d) 0
6 a) $\dfrac{19}{30}$ b) 90
7 a) 60 b) 65 c) 20 d) 80
   e) 100 f) 30 g) 5 h) 20

8 a) C b) B c) A or B
9 a) 0·048 b) 0·0481 c) after
10 Till's bag

Check-up

1 a) depends on your school
   b) certain c) very likely
   d) very unlikely e) very unlikely
   f) very unlikely g) even chance
   h) impossible

2 a) $\dfrac{3}{8}$ b) 0 c) $\dfrac{5}{8}$

3 a) $\dfrac{1}{13}$ b) $\dfrac{1}{52}$ c) $\dfrac{1}{4}$ d) $\dfrac{12}{13}$

   e) $\dfrac{3}{26}$ f) $\dfrac{5}{13}$ g) $\dfrac{3}{4}$ h) $\dfrac{11}{13}$

4 a) $\dfrac{1}{8}$ b) $\dfrac{5}{8}$ c) $\dfrac{1}{2}$ d) $\dfrac{3}{8}$

5 a) $\dfrac{1}{2}$ b) $\dfrac{2}{5}$ c) $\dfrac{2}{3}$

6 $\dfrac{2}{7}$

7 a) $\dfrac{1}{6}$ b) $\dfrac{1}{3}$ c) $\dfrac{2}{11}$ d) 20

8 a) $\dfrac{1}{8}$ b) $\dfrac{3}{16}$ c) $\dfrac{1}{32}$ d) 0

   e) $\dfrac{7}{16}$ f) $\dfrac{1}{4}$ g) $\dfrac{3}{4}$ h) $\dfrac{23}{32}$

9 a) $\dfrac{7}{33}$ b) 0 c) $\dfrac{2}{11}$ d) $\dfrac{19}{33}$

10 a) $\dfrac{1}{20}$ b) $\dfrac{1}{50}$ c) $\dfrac{3}{25}$ d) $\dfrac{1}{4}$

    e) $\dfrac{1}{20}$ f) $\dfrac{3}{5}$ g) $\dfrac{1}{25}$ h) $\dfrac{73}{100}$

11 a) 2 b) 40 c) 26 d) 8
    e) 12 f) 40 g) 24 h) 4
12 a) 5 b) 10 c) 30 d) 15
    e) 15 f) 25 g) 55 h) 15
13 Green marble